Notes from the Energy Underground

Notes from the Energy Underground

Collected by
Malcolm Wells

 **VAN NOSTRAND REINHOLD COMPANY**
New York   Cincinnati   Toronto   London   Melbourne

Copyright © 1980 by Malcolm Wells
Library of Congress Catalog Card Number
 79-24647
ISBN 0-442-25697-3

Printed in the United States of America.
Designed by Loudan Enterprises

Published by Van Nostrand Reinhold Company
A division of Litton Educational Publishing, Inc.
135 West 50th Street, New York, NY 10020, U.S.A.

Van Nostrand Reinhold Limited
1410 Birchmount Road
Scarborough, Ontario M1P 2E7, Canada

Van Nostrand Reinhold Australia Pty. Ltd.
17 Queen Street
Mitcham, Victoria 3132, Australia

Van Nostrand Reinhold Company Limited
Molly Millars Lane
Wokingham, Berkshire, England

16 15 14 13 12 11 10 9 8 7 6 5 4 3 2 1

Library of Congress Cataloging in Publication Data
Main entry under title:

Notes from the energy underground.

Collected by M. Wells.
Includes index.
 1. Renewable energy sources—Addresses, essays,
lectures. 2. Energy conservation—Addresses, essays,
lectures. 3. Human ecology—Addresses, essays, lectures.
I. Wells, Malcolm.
TJ163.24.N67   333.7   79-24647
ISBN 0-442-25697-3

Portions of this book have appeared in slightly different form in previously published works. Grateful acknowledgment is given for permission to reprint the following:

from *Energy Essays*, reprinted with permission from Edmund Scientific Co., Barrington, N.J.: *The Indestructible* by Issac Asimov / *Parkinson's Principle of Pyramiding Pride* by William A. Shurcliff / *Sun Rights; The Window in the Winter; Boom!; Blood, Sweat, And Tears. And Spit; We'll Have the 5-Day Forecast After This Brief Message . . .; Winter; Wilt; Nature, Naked Nature; Antipode; To Tell the Truth; Yecch!; The P.V.H. Indictments; J. F. Mamjjasond; The Birds of Bellazon; Fear of Lying; Waterbags; Foonmanship; Drip. Splash. Creak. Groan.; "Turn a Light On; You'll Ruin Your Eyes!"; Energy in a Bucket; The Topsoil Wars* by Malcolm Wells.

from *Discover, the Philadelphia Sunday Bulletin Magazine.* Copyright 1974/1975, Bulletin Co.: *How Power Became Energy; Sun Rights; The Window in the Winter; Boom!; Heating Up; Blood, Sweat, And Tears. And Spit; Winter; Wilt; Red Buttons; Nature, Naked Nature; Antipode; To Tell the Truth; Yecch!; The P.V.H. Indictments; J. F. Mamjjasond; Waterbags; Foonmanship; "Turn a Light On; You'll Ruin Your Eyes!"; Energy in a Bucket; The Topsoil Wars; What a Way to Go!* by Malcolm Wells.

from *Solar Age*, Church Hill, Harrisville, N.H.: *Nuclear and Solar Economics: A Paradigm Shift Is in Progress* by Hazel Henderson / *59 Ways to Save Energy* by Bruce Anderson;

from *Technology Review* (MIT) and *The Cape Codder: Technological Versus Natural Destruction* by Robert Finch. Reprinted by permission of the author.

from *Commercial Fish Farmer: Looking for a New Aquaculture* by John H. Todd. Reprinted by permission of the author.

We'll Have the 5-Day Forecast After This Brief Message by Malcolm Wells originally appeared in *Harper's Magazine.* Copyright 1973 by Harper's Magazine. Reprinted from the March issue by special permission.

Waste Made America What It Is by Russell Baker originally appeared in *The New York Times.* © 1978 by The New York Times Company. Reprinted by permission.

The Indestructible by Isaac Asimov originally appeared in *The American Way.* Reprinted by permission.

The Flesh Is Always Willing, So Harness It by George Sheehan. Reprinted by permission of George Sheehan.

Contents

I asked all my favorite energy thinkers to contribute to this book, and every one of them came through with an offering. My list of favorites includes my name too, of course, so you'll see it on a disproportionate number of these pages, but the most important ideas are those in the other notes. I hope you'll be as impressed with them as I am.

—Malcolm Wells

Sky Mining

By Malcolm Wells

From the dawn of history until this very moment, man has increasingly built on the living surface of the land for shelter and extracted irreplaceable minerals, fuels, and other resources from *below* it. The consequences of these upside-down activities can be seen outside every window, in every newspaper. We kill the land and mine the earth instead of living *in* the earth and mining the sky.

But even before the great winter of 1977, solar heating and underground architecture had begun to move out of the nutty-ideas category, to introduce a new kind of mining—sky mining, the practice of building into the sheltering earth to use sunlight, wind, rain, and oxygen without the disastrous consequences so common to conventional mining. Earth-covered buildings and solar heating are a perfect combination; an underground solar-heated house in New Jersey needed no fuel at all its first winter!

As an underground architect, I have seen a trickle of interest turn into a river, then into a flood. People all over America seem frantic for information on this subject, because they've discovered that "underground" no longer means damp and dark. It means dry, sunny, and easy to heat.

Here in my underground office as this is being written, the temperature is a steady and delightful 68 degrees. The woodstove's one log a day is keeping the place warm. Outside, in the sunny, snow-covered courtyard, I can see six kinds of birds debating the question of seed-rights. On the roof, under a blanket of snow, pageants of wildflowers are dreaming about next season's most fashionable new colors. And, if I listen carefully, I can hear a distant hum, the rush-hour traffic on a six-lane freeway twenty-five feet from where I sit. Thus do I suffer through the world energy-and-resources crisis.

High living with low impact, that's the promise of the new architecture, but, oh, my, is it slow in coming! Thousands of us may now be building this way, but hundreds of millions of us still aren't. I can't for the life of me understand why so many of us are so eager to build more of the junk that lines our streets, why we care more about trendy styles than we do about what happens when the fuel runs out. Must each new ripple in the energy crisis forever threaten to swamp us? Will we always feel cheated when the power fails? It's insane, and yet our best-known architects continue to build wasteful glass boxes standing in lifeless parking lots.

We want the best of both worlds. That's the trouble. We want glassy wastefulness in architecture *and* free heat from the sun, endless parking lots *and* flowering cities, "high-performance" cars *and* smokeless skies. And of course it just can't be. We must start all over again, right back at what Henry Thoreau called the necessaries of life—the grossest groceries—food, air, water, and shelter, and then go on from there, not exactly back to nature, but back for a closer look at nature.

Everything we can do today with efficient structures, recycling, solar-energy storage, food production, and water purification, for example, is probably done better by the green world. You name it, from computers to pest control, chances are it was perfected by nature before our race appeared. From nature we can learn to build cities that never run out of fuel, silent cities that run on sunlight and wind and rain, that recycle all their wastes and grow many of their own foods, and we can do it not by tearing down the cities we already have, but by unmaking, piece by piece, the mess we've made there, piece by piece. Imagine the number of jobs involved in the unmessing of America. Sky mining is no threat to the multinationals, nor to the energy conglomerates. Many of the biggest ones, recognizing all this, have quietly begun to look for a more appropriate technology already.

It's all so simple, in a way. Here is the living, green world. Here is the free energy of sunlight. Here is the topsoil in which plants grow. So then, how shall we build? Scrape the land bare and pave it with asphalt? Build on top of the warm, sheltering earth and waste the golden sunlight? Of course not, and yet how slowly such perceptions come to us! Everyone is still out to get his share of natural gas or fuel oil before the wells run dry. Very few of us are asking about tomorrow, and the day after tomorrow.

Knowing, as we do, the way cars and radios changed the world, it's difficult to believe they were at first regarded almost as toys, but who in 1915 saw their implications? That's the way it is with sky mining, but if it has even half the potential of the Model T, we're in for some very nice surprises.

How Power Became Energy

By Malcolm Wells

Only a few years ago, the companies that produced electricity were known as *power* companies. They had occasional *power*-failures that resulted in brown-outs or black-outs. And there was even some long-range talk of a future *power* crisis. But it must have been obvious to the power companies, even then, that such talk would remind too many people of the old adage that power corrupts; that absolute power corrupts absolutely. That's how the energy crisis was born; they just changed the name.

Power companies didn't want to leave themselves quite as open to criticism when they switched to nuclear power, so they became energy companies. New Jersey's giant Public Service Electric and Gas Company now calls itself "The Energy People."

"The Power People" would have been more appropriate for a power company, but it might have made us think too much. Apparently we're better off irradiated and happy than safe and thoughtful.

Now they're all on the "energy" bandwagon—gas, oil, coal, electricity. The advertising published by these renamed companies almost convinces me they're the most concerned, selfless, and patriotic of industries. I sleep comfortably in their loving arms. But when I awake I notice that they seldom talk about using much less energy, or about living within our solar budget, or about the ancient rhythms of life we ignore so readily.

When it comes to energy (meaning the kind you have to pay for, not the free kind that shines down upon you all day), the energy companies offer token words about conservation but always follow them with statements about the inevitability of doubled energy-use every decade.

So I look for alternatives. As an architect, I search for ways to build less wastefully, but I often lose sight of my goals and fall back again upon the nonsense I was taught, the nonsense still being taught young architects, that man must improve upon nature. I'm full of that nonsense today, and it takes a new bike-mounted light-generator to reopen my eyes.

Last year, somewhere in Florida, on the leaves of a forgotten sugar-cane plant, a bit of sunlight ended its eight-minute dash to earth. Somehow, the plant turned that sunlight into sugar. Somehow, the sugar got into my sugarbowl, and into my morning coffee. I sipped last year's sunshine at breakfast. Now it's in my blood, and it starts to feed these old architect-muscles. It's dark now, and I start for home. My muscled sunlight suddenly becomes pedal-power, then chain-pull, wheel-spin, generator-whine, filament-heat, and finally—from the headlamp—light again!

Miracle!

In a way, I can hardly believe what I just described. A cynical voice inside me is saying, "The tire spins the generator and the light goes on. Period. No miracles are involved." Maybe. Maybe that's all there is to it, but if that's the case then why is it so damned satisfying? No power companies, no batteries, no nuclear power plants; just a bit of last year's sunlight passing through me on its way back to light again.

The human dynamo pedals his way over the crest of the hill and lets gravity take over. His little light-generator whines away happily, apparently unaware of its changed energy-source. In my mind's eye I see this headlight beam going on forever, slender rays of it racing into the eyes of the oncoming drivers, changing into electrical energy in order to reach their brains, then out into nerve circuits that respond to calls for actions like a steering-wheel correction, a friendly wave, or a muttered curse upon all nighttime bicyclists; my sugared sunlight dissipating itself as heat into the bodies of the oncoming drivers, there to be endlessly transformed into other energy-states by the everlasting restlessness of the universe.

I pedal past a young woman, and the light rays reflected from her many surfaces trigger in me a dozen sensations: Happiness, yearning, sadness, gratitude, self-consciousness . . . all are energy; all sunpower long-since arrived here. "I sing the body electric," said Whitman, who knew something I can appreciate only dimly: it's everywhere, everything, all the time: energy, the most precious thing I know—until I start to wonder where *it* came from. But that's another story.

The green planet with its upturned leaves basks in the warmth of its mother. Sometimes, when I am wise, I know that energy is far more precious, even, than life; not as fragile; eternal; what it's all about: motion, pressure, form, weight, substance . . . all are made of energy; all are equal, all merely parts. I think about tearing up my priorities and rating everything equal to everything else, but that would be going a little too far, so I'll go this far and then stop:

Power—I mean energy—which runs my brain, made me stop and think for a while about energy just now, and saw to it that I was duly impressed.

Generating Power Is No Easy Job!

By Theodore M. Edison

Theodore Edison looks very much like his famous father, but he is no mere shadow of the man. A scientist, inventor, and mathematician in his own right, he has the ability to see straight through complexity to the simplicity at its heart. That and his great sense of humor have made him one of my favorite correspondents.

Power is supplied to modern man in such convenient forms that great waste of it can go almost unnoticed. Perhaps power would be treated with more respect if people could learn to imagine it as muscular effort. To that end, I'm including here some paragraphs from a letter I sent in 1958 to a foundation that was then backing a bullock-power project in India.

In September, 1958, an announcement was made of a "revolutionary [for rural India] new combination water pump-electric generator." It was said that the unit would triple the "amount of water pumped by the ancient Persian wheel (100 gals. a min.), used in Asia for centuries." The announcement went on to say that the generator would produce 4.5 kilowatts and that (pending further tests) the power to run it would be supplied by eight bullocks. In the hope of preventing undue suffering on the part of the bullocks, I sent the following comments to the project sponsors:

I am taking the liberty of writing to you on this subject because the power-per-bullock mentioned seems to me to be very high. The reason that I have the audacity to send you a letter of this kind is that many years ago I took part in building an exhibit that served to illustrate the amount of energy stored in a slice of bread, and in the course of work on that exhibit, I gained a very vivid impression of the significance of "watts" in terms of muscular exertion. In this mechanical age, in which tremendous power often comes in small packages, I doubt that many people appreciate what a truly great amount of effort is required to produce one horsepower (which is about the amount of power each of the eight bullocks would have to produce to generate 4.5 kw—assuming 75 percent generator efficiency).

In testing our exhibit, we found that a very strong man in good condition could work at a 1-horsepower rate (about 0.8 hp useful output) for a *few seconds*—but I personally found that trying to *maintain* even one-tenth of that power output became extremely tiring after only a few minutes. My fear is that establishing a quota of 1 hp per bullock (if that figure is used) may not only subject the bullocks to "torture of the damned," but also make the units *eventually* uneconomic by shortening the lives of the animals. Several considerations lead to this fear, as follows:

(1) On the basis of relative size, I should think it would be fair to take 10 to 1 as a reasonable ratio for the power outputs of bullock versus man—especially as the strength of living things seems to increase much more slowly than the size. Assume that 0.1 hp for man equals 1 hp for bullock. A 150-pound man would have to climb only 22 feet in 1 minute to work at a 0.1-hp rate—and that may not seem like much, but that same rate would take him to a height of 2 miles in 8 hours. I'd hate to face the prospect of climbing a ladder 2 miles high every day at a more or less steady pace! The task would be equivalent to climbing from the ground to the top of an 18-story building every 10 minutes throughout each day (48 times in 8 hours).

(2) If the new unit is to triple the amount of water pumped (per bullock), I wonder if each bullock will not have to work substantially three times as hard as before—and I wonder if the old rate was not established as an upper limit of endurance on the basis of countless years of experience, taking into account hot climates, poor food, and other adverse conditions. I strongly doubt that capacity can be increased that much without *greatly* increasing the work load, because many old devices are not as inefficient as they may at first appear to be. (If an old grandfather clock were as *inefficient* as a modern electric clock drawing only 2 watts, the weekly wind-up of a 10-pound weight would be more than 14 miles, instead of only 4 feet!) Animals may not be well treated in underdeveloped countries, and among the adverse conditions mentioned

may be poor health, sores under yokes at the very points where further work would lead to maximum suffering, etc.

(3) When bullocks turn the old Persian water wheel, the driver gets at least a visual picture of the work being done, whereas he would get no such picture when bullocks drive an electric generator (or even a centrifugal pump). In fact, modern machinery, in which torque may increase with speed, may give a *false* picture by turning easily at slow speeds. With the old water wheel, output is approximately proportional to speed, but with modern equipment, a square law may be involved, and in that case (for example), a 5-fold increase in speed might necessitate a 25-fold increase in power input. Lack of understanding of that factor could lead drivers to underestimate what they were requiring of their animals, and this in turn could lead to mistreatment.

(4) Unless batteries, or other "reservoirs," are used in connection with electric generators, it would seem reasonable to suppose that power outputs would be expected to be rather steady. Elimination of rest periods would greatly increase the exhausting effect of even light work loads.

(5) Work that places continuous tension on some part of the body, or that restricts activities to a limited group of movements for long periods of time, is tiring in itself, no matter how small the power output. Soldiers often faint just from standing at attention. Perhaps I am particularly sensitive to tensions, because I have arthritis and acquire severe "aches" when I work too long at one task. (If arthritis results from tension and worry over what the future may bring, I should think all bullocks would have it!)

(6) If the equipment specified should eventually prove to have an unrealistically large capacity, it might have to be run at less than optimum output. That would reduce efficiency and lead to much waste of both money and bullock effort.

It is so hard to gain an appreciation of the effort required to light a 60-watt lamp on a *sustained* basis (via a 75-percent-efficient generator) that I seriously suggest the following *personal* experiment to those working on the specifications of the apparatus: Load a "sled" with enough weight to require a 20-pound drawbar pull to *keep* the sled in motion; then drag the sled along steadily at about 2 miles per hour for at least half an hour. (Here, the power input would be about 0.107 hp = 80 watts, which may be the approximate "human equivalent" of 1 hp per bullock.)

I regret to say that I don't know what has become of the bullock power project, but I hope that my comments on it will aid the cause of energy conservation by making people more conscious of any power they may be wasting.

The Indestructible

By Isaac Asimov

One of the happiest discoveries of my life made me burst out laughing at 37,000 feet. I was on an American Airlines flight to Dallas, reading the flight magazine, when the punch-line of this Isaac Asimov article caused my outburst.

We hear so much gloomy talk about the energy crisis and our austere future it's nice to know that one of the real experts sees things differently. See if it doesn't make you happy, too.

Some of the most impressive changes of the last century have involved the vehicles by which human beings have been entertained. We went from player pianos to record players; from vaudeville to motion pictures; from radio to television. We added sound to movies, images to radio, color to both. And there seems no doubt we can go further still.

By the use of laser beams and holography, we can produce three-dimensional images more detailed than anything that can be done by ordinary photography on a flat surface. By modern taping procedures we can produce TV cassettes on any subject, so that the individual can play what he wants on his own set at a time that suits his convenience.

Every new advance outmodes the older devices as people flock to the technique that gives them more. The motion picture killed vaudeville, and color killed black and white. No doubt three dimensions will kill flatness and cassettes may kill mass-produced general-purpose television.

And toward what is it all tending? What will be the ultimate?

I watched a demonstration of TV cassettes once and I couldn't help but notice the bulky and expensive auxiliary equipment required to decode the tape, put sound through a speaker and images on a screen. Surely the direction for improvement will be that of miniaturization and sophistication, the same process that in recent decades has given us smaller and more compact radios, cameras, computers, and satellites.

We can expect the auxiliary equipment to shrink and, eventually, to disappear. The cassette will become self-contained and will not only hold the tape but all the mechanisms for producing sound and image as well.

With miniaturization, the cassette machine should become more and more nearly portable; eventually light enough, perhaps, to be carried under one's arm. It should also require less and less energy to operate, reaching an ideal where it would require no energy.

A cassette as ordinarily viewed makes sounds and casts light. That is its purpose, of course, but must sound and light obtrude on others who are not involved or interested? The ideal cassette would be visible and audible only to the person using it.

Present-day cassette machines involve controls, of course. There must be an on-off knob or switch, as well as devices to regulate color, volume, brightness, contrast, and all that sort of thing. Naturally, the direction of change will be toward simplification of controls. Eventually, there will be a single knob—or perhaps none at all.

We could imagine a cassette machine that is always in perfect adjustment; that starts automatically when you look at it; that stops automatically when you cease to look at it; that can play forward or backward, quickly or slowly, by skips or with repetitions, entirely at your pleasure.

Surely, that's the ultimate dream device—a cassette machine that may deal with any of an infinite number of subjects, fictional or non-fictional, that is self-contained, portable, non-energy-consuming, perfectly private, and largely under the control of the will.

Must this remain only a dream? Can we expect to have such a cassette machine someday?

The answer is a definite, "Yes!" We will not only have such a cassette machine someday, we have it now; and we not only have it now, we have had it for many centuries. The ideal I have described is the printed word, the book, the magazine, the object you now hold—light, private, and manipulable at will.

Does it seem to you that the book, unlike the cassette I have been describing, does not produce sound and images? It certainly does.

You cannot read without hearing the words in your mind and seeing the images to which they give rise. In fact, they are *your* sounds and images, not those invented for you by others, and are therefore better.

All media of amusement other than the printed word present you with a pre-packaged image, or sound or both, in greater and greater detail as technology improves. The result is that the media demand less and less of you. Even musical cues and laugh tracks are provided so that particular emotions will be drawn out of you without effort on your part. If reading is difficult for a person (and it is for most), it is to this pre-packaging he or she will turn, and a passive spectator he or she will be.

The printed word presents minimum information, however. Everything but that minimum must be provided by the reader—the intonation of words, the expressions on faces, the actions, the scenery, the background, must all be drawn out of that long line of black-on-white symbols. The book is a shared endeavor between the writer and the reader as no other form of communication can be.

If you are, then, of that small and fortunate minority for whom reading is easy and pleasurable, the book, in all its manifestations, is irreplaceable and indestructible, for it demands that you participate. And, however pleasurable spectatorship may be, participation is better.

Reading isn't the only thing that's easy and pleasurable for Dr. Asimov. He is the author of 161 books, including Science Past–Science Future, *published by Doubleday.*

Parkinson's Principle of Pyramiding Pride

By William A. Shurcliff

William A. Shurcliff is a Harvard physicist and a solar consultant who publishes books on solar-heated structures. I am a confirmed Shurcliff fan.

Every lone-wolf inventor—but no big-corporation executive—knows that when the U.S. Government hands over $500,000 to a big corporation to develop a reliable and cheap solar-heating system, the corporation struggles valiantly but produces a system that is hopelessly complicated and expensive.

Contrariwise, if a lone-wolf inventor, using his own savings, starts to design and build a solar-heating system for his own house, he has a good chance of ending up with a system that does the job and costs little.

Bewildered by this paradox, I consulted economists, engineers, and corporation-heads. But to no avail.

Finally, I consulted the famous Parkinson—the man who discovered that, nine times out of ten, when you enter a crowded post office and carefully choose the shortest line to wait in, *any other* choice would have been preferable.

Parkinson listened to my question, gazed at the ceiling for a minute or two, and said,

The explanation is simple enough. When you pay a corporation $500,000 to invent a solar-heating system, the corporation officials have no choice at all: there is no way that they can escape making a very complicated, much-too-expensive system. They are 100 percent trapped by the Principle of Pyramiding Pride.

They order all their department heads to cooperate fully. Each department head enlists the aid of his best engineers. Each engineer tries to design the very best equipment.

The public relations director calls a press conference and explains that the company is about to score the long-needed breakthrough in solar heating; is about to Lead the Way out of the energy crisis.

Hearing this, the engineers redouble their efforts to devise the very best equipment, so everyone will agree that the Government has gotten its money's worth—NSF officials will be wreathed in smiles, the technical journals will publish salvos of praise, and the company stockholders will cherish their feelings of reverence for the company.

The project is completed on schedule. The result is a technological masterpiece! A joy to behold! It is praised by everyone: Government officials, company heads, press, and public. (The *user* soon finds the equipment to be too complicated and costly to maintain, and—quietly, shamefully—he abandons it.)

I was shocked. "This can't be true," I pleaded, "surely the corporation-heads are intelligent. They know the meanings of the words 'reliable' and 'cheap'."

"But they have no choice!" Parkinson replied. "Surely you remember the Central Africa writing-machine competition?" I did not, and he explained:

Fifteen years ago the United Nations awarded identical contracts to two corporations, Trans-World Products, and the Sam Botts Co. Each was given $1,000,000 and told to design a writing machine that would be truly suited to African countries; the device was to be capable of writing in small letters or large, in English, French, German, or Swahili. It was to withstand tropical dampness and floods.

The Trans-World Products engineers went to work with a will. They used up all the time and money allowed. They produced a 200-pound stainless-steel machine, housed in a fiberglass container which included a rechargeable battery, a 5-year dessicant cartridge, flotation gear, and a 100-page maintenance manual written in twelve languages. Although the first model cost over $100,000 to build, later units could be mass-produced, it was claimed, for only $1,500. The device was a marvel to behold, and the world was lavish with its praise. The president of the company was given a 15 percent salary increase, the department heads were given bigger offices. Even the stockholders in the company felt ennobled by being involved in such a successful and altruistic project. TWP's final report (in four volumes, and weighing eight pounds) is available in all major libraries.

The Sam Botts Co. took no visible action for many months. Old man Botts said nothing to his department heads. He asked no one for help. He built nothing. Day after day he sat in his small office staring off into space. Finally, he mailed off a small package (a Manila envelope which required 20¢ postage) to the sponsoring agency. The envelope contained an ordinary Faber Co. wooden pencil, a check for $990,000 and a brief note which read: "This machine—a pencil—meets the requirements; it writes in any language, is unaffected by damp climates, and, when caught in a flood, floats. Am returning the money we didn't need. Yours truly, S. Botts."

The sponsoring agency was furious with Botts. The press ridiculed him. The stockholders felt crushed; they cut his salary and eventually eased him out of the company entirely.

Today, there are 3,237,000,000 wooden pencils in use in Africa. No second TWP machine was ever built.

Before I could catch my breath, Parkinson said, "Sorry. Must leave. Am to attend the dedication of the new NSF-funded solar-heating system of the local high school. They say it is a technological masterpiece!

Thoughts on Energy

By John Holt

A couple of years ago I flew all the way from Philadelphia to Charlotte without knowing it. I was reading John Holt for the first time, absorbed, watching him shoot holes in some space-colony proposals. His aim was so good I went right out and bought all his books, most of which are about the way we miseducate our kids. John currently edits a newsletter called Growing without School.

One thing more people could do to save energy, even people like me who live in a city apartment and have little or almost no space in it to grow food, is to eat more, as much as they can, of organically raised food. Most people think of this as a matter of health rather than energy, but it has a great deal to do with energy. It now takes 14 calories of energy to produce 1 calorie of food energy, and this figure is rising every year because of the increased reliance on increasingly expensive machinery, fuels, pesticides, fertilizers, and so forth.

It is worth mentioning, however, that the difference in price between organically and non-organically raised food is in many cases steadily shrinking. A few years ago I did not buy organic eggs, because I thought I could not afford to. They were at least half again as expensive as non-organic eggs. Now I can buy organically raised eggs from free-running chickens for exactly the same price that I can get conventional eggs in my local grocery store. But even if organic food still costs us a little more, by buying it and eating it we are in some small degree reducing the amount of energy needed to grow food, helping to maintain or restore the fertility of our land, and giving some encouragement and livelihood to people doing that kind of work. It is a small step, but one that a lot of people can take.

About seven or so years ago, as I write, I sold my last car, and have not owned one since or had any plans to own one. Of course, this is easier in Boston than in most places; it is a compact city, with a public transportation system that will take you to many places. But I walk to a great many places. I suspect we would find, if there were some way to make such a survey, that the average urban journey of a mile or less—between any one point and any other—can be covered about as quickly on foot as by any other means of transportation, if we include the time spent looking for a parking place, getting the car into or out of a garage, etc. A person on a bicycle will beat an automobile driver on most urban journeys of anything under about ten miles. It seems to me we need something intermediate. I would like some very small, portable piece of transportation which would enable me, without much effort, to move about twice as fast as I can walk. I think I know what such a device might be.

Suppose we took something like a skateboard, and instead of mounting it on small wheels, mounted it on quite large wheels with roller bearings, the foot platform suspended beneath the axles. I think we would find that, except in the most hilly cities, we could hum along quite nicely at six to eight mph, and then when we got where we were going, we could simply pick up the gadget and take it inside with us. If anyone reading this wants to design such a gadget, be my guest. All I want to know is how it comes out, and perhaps have a sample if you get around to making one.

I won't add to what has been said about solar or wind energy, because that is all discussed so often. Let me take up something different. About seven months or so ago I started raising worms, the regular commercial wrigglers, in a wastebasket in the kitchen of my basement apartment. I bought them from one of the big worm-supply firms; they arrived in good condition. I started them off on the peatmoss that was sent with them. Earlier in the fall when the Boston Public Garden was full of leaves, I carted about eight loads, or 250 pounds, of these to the little sunken patio outside my kitchen door. All during the late fall and winter, and right up to now, I have saved just about all of my gray water, and I dump this water, frequently laced with urine, on these leaves. The leaves have partially decomposed and make a food that the worms just love to eat. Later I added some grass clippings from the Public Garden, but I have also learned, just as the books said, that the worms love cardboard, the regular kind that boxes are made of. I cut cardboard into strips with a paper guillotine (this may not be necessary, as far as I know), and the worms just eat it up.

This saves energy for several reasons. It costs energy for cities to take their solid wastes from the city to somewhere else. By recycling these wastes and turning them into a rich topsoil—for which I can surely find many uses—I save some of this energy. I also cut down on the pollution of our waters by recycling my gray water in this manner, which again in the long run must conserve energy. I know that worms eat all kinds of animal manure, but I have not yet taken the time to try out on them either my own or the dog's droppings, which are so plentiful in my part of the city. I have seen worms eat rabbit and other kinds of manure, which they gobble up directly, but it may be that the stronger kinds of feces that we or other carnivorous animals produce might have to be mixed with some crushed limestone, sawdust, or something like that to make it a little less strong. But by a year from now, when I will have a great many more worms than I have now, I will probably try this out. I hope in time to be able to take myself out of the waste cycle altogether.

I am very interested in the possibility of people in cities, including poor people, raising much of their own food. A good deal of work has been done on this already. People in a number of cities, and as far north as Montreal, are raising a considerable amount of food in rooftop gardens. Other people have had some success raising fish indoors in small tanks. Still others raise sprouts. I read not long ago in *Mother Earth News* that the San Diego Zoo, in a fairly small room, grew and harvested 500 pounds of sprouts a day to feed its animals, and at a cost much less than the cost of baled hay. Surely here is a way in which a great many city people, even the poorest, could raise much healthful food, and in so doing cut down their food budgets and have more money for other things.

I am also told that worm castings make a good medium for mushroom growing.

Some people who read this piece will probably know about the concept of Net Energy, an important idea conceived by Howard Odum. Briefly the idea is this. If it takes 100 units (of whatever size) to produce some kind of energy-producing or -converting device, whether it is an oil well, windmill, solar heater, etc., and that device during its lifetime will produce 150 of the same units of energy, then the net energy of that device is 50 units. So to the concept of Net Energy I would add the concept of Net Energy Ratio or Net Energy Factor, which is the net energy of the device divided by the energy cost of the device. Thus, if something took 100 units of energy to build, and if it converted or provided us with 200 units, it would have a net energy of 100 and a net energy factor of 2.

The point of all this is that our energy problems, however we define them, are not going to be "solved" by things like nuclear reactors or even thermonuclear reactors (if we could ever develop them, which believe me is a *long* way away), which have very low net energy factors, or even negative factors. There are very good reasons to believe that if we calculated *all* the energy costs of present-day nuclear reactors, their net energy would be minus. One of the reasons that nuclear energy is made so attractive to many people is that the people who push it are very good at externalizing or hiding many of its energy costs. Thus, we have not yet started to take into account the energy costs of disposing of wastes, dismantling reactors, etc., because we have not even found out how to do it.

I have said that the fusion or thermonuclear reactor, even if we could build it, would not solve our energy problems, as everyone seems to think it would. The reason for this is quite simple, and can be put in the form of a question: How much would you pay for a perpetual motion machine, or a car that ran on no fuel, or fresh water, or something like that? Many people might be willing to pay $20,000 for a car that would never require any fuel. But it would be a complete waste of money to pay $100,000 for such a car, because we would never save enough money on fuel to cover the cost of the car.

The energy costs of building and maintaining a fusion reactor will be simply staggering. For one thing, they will include all the energy costs of developing it. An engineer who works on one of these projects, which by the way is a long way from being large enough to sustain a thermonuclear reaction, told me that in order to get their gadget going they have to supply it with an amount of electricity equal to the total amount used by the city of San Diego, with over a million people. To supply this, they had to build a special generator, the armature of which alone weighed about 15 tons, and at full power revolved at 500 rpms. He said that in order to get this generator up to full speed, they had to run a 2,800-hp electric motor—quite a piece of machinery in itself—for five minutes. When this machine was turned on, the drain on this enormous generator was such that the drain reduced the generator's speed by one third. To run this machine, they had to use, among other things, the largest electric switch in the world, which was made in Japan and imported to this country. The engineer described an accident that had taken place. Something broke the flow of current, and the currents were so enormous and the collapsing magnetic field so powerful that they tore huge machines out of their mounts, pulled large bolts out of the metal they were bolted to, and so on. There will be no small accidents in thermonuclear reactors. We may not have the kind of disasters that are possible with fission reactors, but we may be sure that when something goes wrong with a fusion reactor, fixing it will not be a matter of a couple of days.

There will be a big outcry of publicity when the fusion reactor is delivering power. But it is liable to be a very long time before any such reactor will get its net energy even up to zero, and a still longer time before any such reactor will be able to produce enough energy to build even one other similar reactor. In other words, for a very long time thermonuclear reactors are likely to be a drain on conventional energy sources rather than a supplement to them.

Water

By Steve Baer

"Sorry," wrote Steve Baer when I asked him for some biographical material, "no things I'd like to have known about myself. Hell, for a piece as short as this what do you need, anyway?" When you're the guru/inventor/philosopher of the solar movement, you don't need an introduction. I just hope this Baer fragment will lead you to his other works.

What if someday someone rediscovers water—just as today many people have rediscovered solar energy. It will seem an impossible task to many to get people to drink water. How could it compete with soft drinks, beer, wine, Kool Aid? Where could you find water? There is no water industry. Its source, the rain, will continue to be overlooked. Finally the government will turn to the Coca-Cola Company to put it out as a new drink—Water (by Coke).

The Solar Buick

By J. Baldwin

J. Baldwin asked me to "make it clear that (his) article was not a carefully written, edited thing, but a personal letter to (me) as is." A more-with-less man, he can build, fix, design, or explain anything, and I'm delighted he has left California and has come to Cape Cod, even if it's only for a little while. J. is also a musician, a whitewater river guide, and a writer.

Take a 1953 Buick Roadmaster "four holer," remove the vast straight-eight engine and replace it with a Briggs & Stratton. Do you now have a recycled, charming automobile suitable for these times of environmental concern? I think not. This is because the Buick is both symbol and product of an *attitude* marked by thoughtless, flaunted waste on the part of both manufacturer and buyer. At the time it was built, the waste was disguised as "prestige" and "luxurious comfort," a combination most of us still find difficult to shun. This attitude tends to make clear thought difficult. The '53 Roadmaster also featured truly dangerous handling and braking characteristics, which everyone apparently ignored as thoroughly as they did the abysmal gas mileage. Buickness is still with us, and it is not confined to automobiles; most housing above the mere subsistence level is designed and used as if there were an endless supply of everything.

Retrofitting old, charming Buick houses with expensive copper collectors is about as useful as retrofitting the above-mentioned Briggs & Stratton in Buick cars. I shudder at the government's ill-considered tax breaks and other incentives aimed at such nonsense. The benefits of such programs go mostly to the already-well-off homeowner. Besides, there is a sinister side effect to the governmental meddling (as always). The effect stems from the inevitable tendency for largesse to be lavished only upon "approved" hardware. We can't expect taxpayer's money to be used on ineffective equipment, now can we? I fear that "approved" will soon mean products from "established, responsible" firms like Westinghouse, Grumman, Honeywell, and others of similar stripe. This approval will extend by precedent to FHA loans and thence to everyday bank loans. It will become code. The small pioneering companies that do not grace the Big Board with their tiny finances will be forced out. We will have to buy the sun after all, at a price the market will bear.

Discouraging as that may sound, there is something worse. This is the widespread belief that industrial societies can somehow continue doing the Buick on renewable resources (or "unlimited fusion power"), and that if this can be done it is not only OK, but also desirable. For instance, there are many who should know better, basking in the belief that their woodburning Buick house is somehow exempt from considerations of efficiency and conservation. Again, it is the *attitude* that bothers me, not as much the actual hardware in question.

I recently tangled with a well-known alternative-technology personality who insisted that in Oregon where he lives there is no need to be concerned about having a big, sprawling house and burning lots of wood to keep it warm. "Oregon will never run out of trees," he said. He said that, even though he admitted that he neither planted trees nor managed a woodlot. I have heard that "will never run out" before, haven't you? What's the answer to this type of blindness? What hope is there if those on the side of earth-stewardship see the forests as eternal and go so far as to build their houses from *Redwood* yet? How are we going to raise consciousness if environmentalists think that 3,000-square-foot homes with solar hot-water heaters and diesel Mercedes are an answer or represent a demonstration of benign technology? Alas, and woe is us if this is the best we can do!

Let me whap that *we* a moment. *We*, so far, have mostly tended to make things as right as we can manage psychologically or financially for ourselves and perhaps a few friends. Many of us, including me, have a sordid past, which we mined for the money to join the bourgeois flight to the country or at least escape the 9-to-5 grind. We feather our nests. Some of us make a lot of righteous noise about how we'd like things to be, but few of us go farther than our own family (however defined) or small community. We have not done much as a we.

At a recent workshop attended by government bigwigs and various industrialists, a man from the Department of Commerce observed that we (the righteous ones) "had no credibility." He was right. I was upset. It was such a simple, deadly truth, the sort of thing my Daddy was good at stating. The government man said it as a Daddy, too. Damnation! So how do we become credible?

Several thousand scattered country cottages with real Swedish woodstoves, windmills, goats, and '63 Chevy Pickups (each of which needs yet another crack in the head repaired) don't constitute credibility in anyone's book, not even ours. What to do? And one should also remember that the bestselling automobile in this country is still the Oldsmobile, which is only a wee bit less gross than a Buick. . . We are still very definitely a minority.

I think that what is needed is a significant number of people, several thousand at least as a start, who live a good life free of bizarre gimmicks, yet equally free of Buick overtones. I am proposing a demonstration of earth stewardship at a scale that would be irrefutable. I think it can be done without giving up much of what we consider to be "necessities" or things that we are very used to. We need only give up the waste.

There have been several studies showing that our society is mechanically only about 5-percent efficient. Even the best aerodynamic cars are only about 10-percent efficient, and most are more like 6 percent. Such figures are easily flipped about, but consider for a moment what that means. It means for every 100 gallons of gas you buy, you *throw away* 94 *gallons* in the form of pollution of one sort or another! That's hard to believe, isn't it? But it's true! An incandescent light bulb at 4-percent efficiency ends up overall, from coal mine to light, being way less than 1-percent efficient. This is because efficiency percentages multiply. For instance, a 4-percent lightbulb fed by a generator that's 40-percent efficient is, overall, 1.6-percent efficient, and that doesn't count all the efficiency losses in the wires or the percent of the mined coal used to operate the coal mine! In a word: shameful! For this we are going to nukes and strip mining!

I have elected to work on technological efficiency as my gig, on the theory that there are many people who would like to do better, but the hardware isn't there to let them. I'm betting that there are a great many people who would do better if they could.

I'm betting that there are thousands of people who would like to try a community based on earth-stewardship, a community with technology designed to do a lot better than that 5-percent efficiency. Passive solar buildings are an example of what I mean. In many there aren't really any mechanisms, yet the houses are warm or cool as required. To an engineer, speaking of this in terms of efficiency is not quite textbook thinking, but to the people in the house it means comfort they are used to, and fewer scarce resources used. Of course I suppose one *could* make a passive-solar Buick house, but the costs of a house rise drastically as the size goes up, and I would expect there would be social pressures to discourage such extremes. How many people would feel that excessive size bestowed prestige, if passersby scorned their palace? Instead, I bet people would take pride and prestige from being conspicuously efficient. They could use the money saved for other things, or could maybe work only half time instead. Go fishing.

So I think that the time has come for a non-fossil-fuel-burning community to be founded, built and lived in by some of *us*. With commitment. Maybe even love. How else will we make any big difference unless we *demonstrate* that it can be done and done well. What better antidote to the Buick than to change the rules of prestige while at the same time actually to increase comfort and security? Uh huh, security. That's what the Buick is really about. We know now that Buick security is ephemeral at best, certainly isn't appropriate, and certainly won't last very long. The security found in the proposed community would be lasting (barring an H bomb) and would be a true security instead of mere appearance of security.

The hardware is about ready, thanks to the New Alchemists' Ark projects. Arks are solar-powered bioshelters made of low cost materials, and housing aquacultures and vegetable gardens. Arks enable a community to be built on cheap land in tough climates, yet raise all its own food. Clever architecture of various proven configurations greatly reduces energy needs and makes it feasible for a community to make its own power. The rest can take many forms and would be what seemed good to those concerned. With the hardware ready, the need acute, and the people up for it, I feel the time has come to do the deed. *We* will do it, who else? Exciting, yes? Let's go!

The Case Against Government Standards for the Solar Industry

By William A. Shurcliff

It seems to me that it would be a great mistake for the government—any government—to establish mandatory solar criteria.

If the material were meant merely to inform manufacturers, or to constitute *suggestions* as to criteria and tests, it would be highly valuable. The suggested criteria and tests could be set forth in a clear and well-organized way.

But to establish these suggested criteria as mandatory, and to deny various benefits to persons buying equipment not satisfying the criteria would, I believe, be a great disservice.

Specifically, mandatory criteria:
- would impose a great burden (expense, worry, argumentation, delay) on the manufacturers who tried to comply.
- would virtually destroy companies that had neither the money nor the time to seek certification, or did not have a legal staff competent to analyze and weigh the thousand or so criteria and tests, including criteria and tests that were sufficiently general, or vague, as to leave a reader in great doubt.
- would tend to inhibit invention (without meaning to) of radically different, and perhaps cheaper, systems. Such systems would not fit into the framework presented, and the burden of proof would fall onto the proponent of an innovative scheme. He would have to decide what constituted proof—and do the proving! In effect, he

would be declared guilty unless he could somehow prove himself innocent.

I believe it may be true that, today, very few reliable solar-heating systems that fit such standards are cost-effective. It just may be that the standards rule out most of the cost-effective systems.

Does any government staff know how to make a truly reliable and cost-effective solar-heating system? If it does not, how could it have the temerity to dictate such standards to others? When Einstein was drafting an article on his theory of relativity, should a committee have attempted to tell him what format to use—what terms, what kind of equations?

Why is there this intense, concentrated government concern that John Q. Public not suffer any disappointment when he buys solar-heating equipment? By way of contrast, consider the following:

If he buys a second-hand car and it turns out very badly, he has little or no recourse.

If he buys cigarettes and gets cancer of the lung (as 100,000 persons do each year), he has no recourse.

If he buys alcoholic beverages and becomes an alcoholic (as 1 million persons do each year), he has no recourse.

Why, then, this tremendous concern that he might be wasting, say, $2,000? Is this as hard on him as buying a very defective $4,000 car, or contracting cancer of the lungs, or becoming an alcoholic? Considering cigarettes, alcoholic beverages, boats, swimming pools, snowmobiles, etc., do not our citizens waste on the order of $10 to $100 billion each year? Is the government really in the business of trying to stop people from wasting money?

How far would Henry Ford have gotten with his Model T if a set of certification criteria had been instituted by the government in, say, 1913? Would not Henry Ford have had to close down his factory if the government had insisted that any car, to be certified, had to have a self-starter, had to have 30,000-mile tires, had to have four-wheel brakes, had to have safety glass, etc.? Even though his Model T would have flunked every such test "by a mile," the car was a tremendous success. It had terrible trouble, but it was such an improvement over the horse and buggy that tens of millions of people flocked to buy the car—including my father. (When I was 16, I drove his Model T, suffered with it, and found it to be, on the whole, an enormous success.)

Those who propose to refuse to certify a solar-heating system that entails introduction (into the house) of bacteria or materials that might cause disease forget that more than 10 million homeowners keep dogs, of which about 1 million are not house-trained and about 5 million have fleas.

Those who would refuse to certify a solar-heating system that cannot be serviced with ordinary tools, but require special tools, overlook the alternatives, such as using up the last of our oil, or burning many millions of additional tons of coal with resulting widespread pollution and damage to millions of people's lungs.

I do not enjoy seeing a person spend money on a poor product. Also, I already know of some solar-heating products that are poor. And I favor honesty in advertising, full disclosure, thorough testing, etc. But to try to impose standards—on a fast-growing, very young, not-really-yet-formed industry—seems to me a great disservice. Just as the imposition of standards could have stopped the automobile industry dead, could have stopped the minicalculator industry, and has notoriously crippled innovation in the building industry, so imposition of standards on this young solar-heating industry seems to be a great mistake. To draw up, to impose standards is, in a small way, to play God.

To impose standards on life-and-death materials like vaccines is essential.

But on solar heating systems, NO. I am no economist, and I may be wrong; but these are my views. I am not trying, really, to criticize anyone's use of detailed solar specifications, but only the broad idea that anyone in government should give such requirements the force of law. By way of analogy, if someone should decree that "all vehicles should have four soundly made wheels," I would criticize this—not because I'm against soundly made wheels, and not because use of four wheels is necessarily bad, but because (unwittingly) such a decree would rule out entirely bicycles, 16-wheel trucks, and 10-wheel transport planes.

Of course, the people establishing the criteria mean well, and their criteria are in some sense tentative and subject to change, and they can allow exceptions. But the overall effect is to crush a small, innovative group that can't afford to hire a $50,000 lawyer and can't afford to wait two years for action on its appeals.

Moods of a Variable Star

By David H. Thompson

*David H. Thompson told me that his experience with the extreme
environments of the south-polar regions led to his present interest in the
environments of outer space and astro-pollution. Could be. But I still
marvel at a mind that can leap from penguins to space junk, and to this.*

If the earth is our mother, then the sun is our father. Without the sun,
there would be no life. This ball of blazing plasma provides most of the
energy used on the earth. It feeds the plants that gobble its rays with teeth
of chlorophyll. The carbohydrates they produce ultimately drive the
whole ecosystem. It will provide energy when our flirtations with limited sources or Faustian technologies turn sour. And now astronomi-
cal studies are showing new and subtle influences of this throbbing orb
upon our lives.

The sun is a variable star, which means that astronomers have to
keep on their toes or they might miss something. In fact, the sun goes
through complex cycles of sunspots. These are not the annoying red
spots seen shortly after one looks at the sun. Rather, a sunspot is a
gigantic, dark-colored solar storm with a diameter of up to 90,000 miles.
One puny sunspot could easily swallow the earth if it had an appetite,
since our planet has a diameter of only 7,900 miles. Sunspots become
more numerous in a cyclical fashion. There is a basic 11-year cycle
between maximum numbers of sunspots, but every other peak is larger,

creating also a 22-year cycle. Similarly, there is an even longer cycle of 80 to 90 years, and there may be still longer ones not yet identified. Other solar activities vary with the sunspot cycle. The next sunspot maximum will occur in 1979–80.

Over the years, everything from the prices of shares and commodities, political events, earthquakes, and even population fluctuations of animals has been correlated with sunspot cycles. Many of these claimed relationships are probably spurious. But there is solid statistical evidence to show that climate varies with both the 22- and the 80- to 90-year solar-activity cycles. For example, the periodic droughts in the western United States are linked to the sunspot cycle, and our solar weatherman predicts a return of dust bowl conditions to the central plains around the turn of the century.

The effect of sunspots on short-term weather, as opposed to long-term climatic change, is more controversial. But recent studies are suggesting that the frequency of severe thunderstorms also varies with sunspots. Since thunderstorms and associated lightning cause considerable damage to electric transmission facilities, the utilities will have to take their heads out of the sand and look at the sun, even if they can't be persuaded to forsake coal or atomic power.

Changes in the sun can affect us in ways other than through the weather. For example, the amount of ultraviolet light from the sun varies substantially as sunspot numbers wax and wane. It can vary by as much as 15 percent over a period of seven years. When moderate levels of ultraviolet light fall on our skin, it helps our bodies manufacture vitamin D, which builds strong bones and thereby prevents fractures. Yes, even our skin is a solar collector of sorts. But severe levels of ultraviolet light irradiating the skin can produce sunburn and skin cancer, and even temporary "snowblindness" when the eyes receive an overdose. Crops and many tiny plants and animals essential to the ecosystem might also be harmed by excessively high ultraviolet levels.

Our fickle sun can influence the ozone layer high in our atmosphere by variations in ultraviolet light and in the solar wind. This in turn may lead to changes in weather or climate, since the amount of ozone influences the energy balance of the atmosphere. The ozone layer also helps to shield us from excess ultraviolet light. Already, English scientists have detected correlations between changes in the protective ozone layer and the numbers of fractures of certain bones. When the ozone is thicker, less ultraviolet light reaches our skin, producing less vitamin D, making the bones of some people more brittle. And a Russian scientist claims to have detected "anomalies" in human blood that correlate well with the timing of a large solar storm in 1972.

There are still other solar impacts on our lives. The tides are mostly due to the action of the moon, but when the moon lines up with the sun, the tides get an extra boost. We are constantly bombarded with extremely powerful particles flung from mysterious and unknown dynamos of the universe. Termed cosmic rays, these penetrating atomic bullets produce the northern lights, trigger some birth defects and cancer, and may even contribute to the flickering pinpoints of light we perceive in an absolutely dark room. The magnetic field of the sun deflects many cosmic rays, but again, this shielding effect changes with the sunspot cycle.

The sun even produces "weather" in space. The "solar wind" is an extremely hot, rarified gas that moves with a speed of about 864,000 miles per hour, taking five days to reach the earth. Moreover, the solar wind is gusty! And periodically, storms in space are produced by eruptions from sunspots and other active regions of the sun. These solar "flares" will send astronauts scurrying for their lead-lined underwear or storm "cellars." During flights to the moon, ground controllers kept a close watch on the sun, ready to abort Apollo missions at the first sign of trouble. Some observatories still watch for the beginnings of a flare, and the National Bureau of Standards' radio station, WWV, daily broadcasts a solar-terrestrial forecast 18 minutes after the hour.

In spite of these facts, some people are trying to sell lots in their "L-5 Industrial Acres" in space. "The lot with the view," they say. One of their hypes is that "there is no weather in space." Real estate agents used to say the same thing about California, until the people who had been swept away by mudslides, floods, or Santa Ana winds, or were parched by drought, objected.

"But," you retort, "this is all very far removed from my everyday existence. It couldn't possibly have any impact on my life."

Skylab almost had an impact. The largest object ever placed in orbit was in trouble. Placed in a low orbit at the fringes of our atmosphere, it was predicted to stay aloft until the Space Shuttle could boost it to a higher orbit or bring it safely down. But sunspot cycles weren't analyzed correctly. NASA's solar forecaster was wrong. More sunspots appeared than were predicted. The upper atmosphere was heated by increased solar activity and reached higher to exert increased drag on the abandoned Skylab. Consequently, Skylab reentered the atmosphere in a spectacular, flaming crash in 1979, before the Space Shuttle could rescue it. It was only because of luck that weighty items, such as film vaults or 2,000-pound flywheels, did not crash down on populated areas of the earth.

No impact on your life, you say? What about that expensive CB radio you just bought? The increasing solar activity may render it difficult to use on many days.

Insurance companies, take heed! Everything from satellite disasters and thunderstorms to cancer or broken bones may be related to solar cycles in a statistical way. Your giant computers are itching to sink their integrated circuits into the latest, updated, solar-activity forecast. Insurance companies make profits from statistical straws in the wind before the rest of us know which way the wind is blowing.

Solar energy experts, pay attention! Changes in solar output just now being detected, weather, cloud cover, wind, and tides all introduce subtle nuances into your equations—and subtlety is the essence of solar energy. Knowledge of the sun will bring many benefits. Part of the answer to long-range weather forecasting may lie in a greater appreciation of the sun's moods.

Scientists who analyzed the effects of sunspots were once called crackpots. But now, better data shows that they may have been right after all. As we become more sun-conscious, newly discovered "moods of the sun" must be reckoned with. Citizens of the coming solar age should be just as well tuned to the nuances of the sun as sailors are to every gust of wind.

Letter from New Jersey

By Doug Kelbaugh

Douglas S. Kelbaugh is a Princeton architect whose award-winning solar buildings have begun to win for him the national recognition he deserves. He's also a rascal; as you will see.

Six years have passed since we decided to build our house here in central New Jersey. It's a solar house, but that's a small part of the story. The first revelation about building a house is that it's expensive—probably double what you would guess. If you want to build something worthy of construction, it will cost considerably more than twice your annual income—the old rule of thumb for budgeting a home. But, perhaps we should be spending more on shelter anyway; after all, shifting expenditures from other items is the only way to reverse the very real decline in the quality of our dwellings. The second revelation is that building a solar house is even more expensive. It's a dilemma. You are caught somewhere between a rock and a hard place.

One solution is to build smaller—lower your expectations. American houses have always been notoriously large by international standards. Europeans have opted for less space and built their dwellings better.

We built a house with approximately 1,800 square feet of living space—150 more than designed, because I staked out the foundations wrong. The house really has more space than we need, but the third bedroom has doubled nicely as an office, and my wife has been able to use our greenhouse in her horticultural business.

Our passive solar system added about $8,000 to the cost of the house and the greenhouse added another $3,000. As the accompanying chart indicates, it now saves about $400 a year, which means probably a 15- to 20-year payback (the figures are all a wager on the future costs of fossil fuels). Not a spectacular investment, about a five-percent return, but not bad. What's the payback on a tennis racket or a pair of shoes?

You can forget uniform temperatures, but so what? Why keep all rooms the same temperature every hour of the day? That's a little like driving your car at the same speed all the time. In the winter our house typically swings on the first floor from 65°F during the day to 55°F at night and from 70°F to 60°F on the second floor. Sometimes a little more, sometimes a little less, but if we didn't have thermometers hanging everywhere, we'd be hard pressed to estimate the swing. We've grown to like a cool house. The temperature is often 60°F or 62°F downstairs, and unless you're in a melancholy mood, it's fine. We could keep the house warmer by turning up the thermostat to our backup furnace. But when you can save so much money by keeping it cool, there's a strong motivation to love it cool. That's the way most of the people of the world live, including several countries with higher per capita income than the United States. The historical abberation is a uniform 70°F environment, not the recent return to lower thermostat settings.

If you live in a passive solar house, you'll probably have higher room-surface temperatures, lower air movement, and higher humidity—especially if you have a solar greenhouse. Three out of four of the classic comfort parameters—humidity, mean radiant temperature, and air speed are superior to conventional heating systems. Only temperature is lower—and that's probably to your health. Also, passive solar homes are quieter because the furnace runs less.

The nicest part of living in a solar house is that our total heating bill was $75 last winter. It was a very severe winter, but a very sunny one. This winter our total bill was also $75. It was almost as cold and considerably cloudier.

The second nicest part is becoming more aware of the climate—both on the micro and the regional scale.

The third nicest part is the uniformity of your children's body temperature. Our five-year-old son has run a steady 98.6°F, with a fever only twice in his four years in a passive solar house. His conventionally heated, runny-nose friends seem to be sick all the time.

The worst part of living in a solar house is the lack of privacy and the endless publicity. It's wonderful at first. All those gapers, all those phone calls. But even a ham will soon tire of the incessant questions and even worse, the compliments. A six- to eight-foot solid fence soon becomes a necessity—one of the hidden costs of solar heating. All solar houses will need them until 1983, when the novelty will wear off.

The most important part of living in a solar house is not efficiency or performance but how you feel about it as your home. It's where you spend much of your time, albeit much of it asleep. Designing a house to fit your lifestyle, your aspirations, and your fantasies is much more important than pulling off a solar *tour de force*.

As for us, we just put a down payment on a mobile home in Cincinnati.

Kelbaugh House
'76–'77 Winter Performance

HEAT LOSS:

5556 Degree days; Design Temp.
65°F inside, 0°F outside.
Based on Storage Temp. degradation
field studies [1]   115,500,000 BTU
Based on ASHRAE methods [2]   136,675,000 BTU

HEAT GAIN:

Miscellaneous:

Occupants	3,250,000 BTU		
Electrical [3]	4,800,000		
Gas appliances [4]	4,750,000		
	12,800,000		
Gas furnace [5]	18,450,000		
Sub total	31,250,000		
Solar		105,425,000	84,250,000
Solar/SF Glass [6]		124,000	99,000
Fuel savings [7]		1405 ccf natural gas	1123 ccf natural gas
Dollar value of savings [8]		$450	$360

---

[1] Comparison of decrease over 12 hours in temperature of the Trombe wall, wall, greenhouse water drums, greenhouse concrete floor, and house floor slab, with the hourly difference between indoor and outdoor temperatures. Hourly heat loss at design temperature is 56,300 BTU.

[2] Standard calculation using U factors, slab perimeter loss, and 2/3rds air change per hour. Hourly heat loss at design temp. is 65,600 BTU.

[3] 1406 kwh.

[4] 105 ccf natural gas burned @ 75% efficiency with 100% capture on cook stove, 60% capture on gas dryer and domestic hot water heater.

[5] 246 ccf natural gas @ 75% efficiency. $75. at local rates.

[6] 850 SF glass.

[7] @ 75% efficiency.

[8] @ 32¢ per ccf.

Some Like It Hot

By Don Prowler

The relation between building design and energy consumption is the subject about which architect Donald Prowler writes and teaches—at the University of Pennsylvania. He's really good. A year or two ago he organized a passive solar energy conference there and did such a good job people by the hundreds had to be turned away.

By late April, Philadelphia is usually under the dizzying influence of spring. Winter coats are headed for the recesses of closets, playgrounds are coming alive with the sound of rubber on asphalt, and those hearty urban dwellers, the ginkgo trees, are once again coming into leaf, outwitting all the salt dumped by the Department of Streets' snow-removal machines. So April seemed an excellent time to move into our new office, even if the central heating system was not quite working and the retrofitted Trombe wall* was not yet in place. The new season would usher in our new location; it was poetic, it was prophetic, but mostly it was cheap.

Of course, as luck would have it, this year the temperature plunged to record lows in late April and early May, coinciding exactly with our first few weeks in the new office. While at first our numbed extremities

---

*A passive solar heating device.

were hard to ignore, we collectively decided to accept the experience for what it was: a curse . . . or an opportunity to learn something about energy and buildings. After all, we knew by then that about one quarter of all the energy used in this country each year goes into our buildings, primarily for heating, cooling, and providing domestic hot water.

Still, at first, there was much grumbling about the temperature, and it *was* chilly. It is one thing to talk about energy conservation and resource allocation, and quite another to live it. We were living it with a vengeance. The office was in a brick townhouse, traditionally considered an energy-conservative type of dwelling, but ours was a ramshackle end unit with three sides exposed to the elements and no internal loads (heat-producing equipment) to heat up the place.

"It must be 40 degrees in here," someone would say, his breath forming puffy clouds of condensation as he spoke.

Now, I knew it was cool, even cold, but this thermal hyperbole was excessive and I decided to counterattack. Pulling out all the stops, I purchased a $4.95 Taylor indoor thermometer and hung it next to our not-yet-functioning thermostat. Now I would be armed with facts, dangerous weapons in anybody's hands. Everyone was forced, at least twice a day, to guess the temperature in the building. It was invariably warmer than most guessed. Some people consistently estimated temperatures five degrees cooler than the actual readings indicated, while others fluctuated in their guesses, and only a few guessed temperatures that were too high. The inescapable fact was that we were accustomed to living in spaces much warmer than the 65 we are all asked to live at during the heating season. Readings in that range were generally characterized as cool. The effects of cooler radiant surfaces on comfort became apparent. During the day, our building was acting like a passive solar building in *reverse*, the massive walls remaining cooler longer than the air, which was being heated by the sun.

Just knowing the temperature, however, coupled with an acute energy consciousness, seemed to make cooler temperatures more acceptable. Readings in the 60–65°F range no longer elicited anguished pleas for the arrival of summer. The thermometer, it turned out, was almost as good as a stove.

We all became much more aware of the movements of the sun as it penetrated different windows and created locally heated zones. The weather report on the office radio generated more interest than any hot news item as a vigil was kept for the true reapperance of spring. There were unquestionable benefits, too. For the first time, it was possible to experience, on a daily basis, the energy flows through walls of masonry, insulation, and wood, with different outside temperatures and with varying amounts of solar energy. An informal log allowed us to observe the performance of the building as one might a test facility.

When energy and resource consumption is distilled down to economic formulas, complete with payback periods, life-cycle costing, and years left to positive savings, as it often is, it loses any quality of immediacy, of urgency, and of just plain fairness. The dollar obscures or erases the hard reality of limited access to resources, but our building was a constant reminder of them.

As the season progresses, the days will, no doubt, get longer, and soon the memory of those nippy days will fade. Perhaps, we will celebrate an annual "Energy Appreciation Day" as a reminder. On that day, we will have an "Energy Fast" and use only natural daylighting and no heating or air-conditioning. Perhaps, when it is warmer, we will have time to consider whether we are living with an omen or an aberration.

Of course, there are those who might say that this is all rather exaggerated, to make a few points, but as soon as I can melt the icicles forming on my nose, I plan to go out and buy another thermometer.

Sun Rights

By Malcolm Wells

I used to amuse myself by imagining what the property lines around building lots would look like if they were extended downward and upward as far as they could go. Downward, the boundaries would converge until they met at the center of the earth, touching the slender tips of all the other properties in the world, making neighbors of us all; and upward, they'd widen out forever until they'd knocked whole galaxies askew as the earth spun on its axis and raced around the sun.

But such daydreams occurred before I knew about solar energy. Nowadays, every hike in the price of energy moves us closer to what seems like the only fuel that makes good sense: sunlight. It's free, dependable, safe, and clean. It doesn't deplete natural resources or get us involved with deadly nuclear power. For literally billions of years, down through the ages of the dinosaurs and the saber-toothed tiger, down all those countless centuries and millenia until the discovery of fire, each year's ration of sunlight was that year's ration of energy. Period.

During 99.9 percent of all the time that life has existed on this earth, in other words, solar energy was the only energy there was. If you include *stored* solar energy (food, coal, oil, gas, and wood) you can say almost the same thing about life today; most of our energy still comes from the sun. Still, we treat sunlight the way we treat good health—as something to overlook until we get in a jam. Which we are now in.

The only trouble with sunlight is that it slants. Unless you live in

Ecuador or in one of a few other places I'm sure you don't live in, all your sunlight comes toward you at an angle. Slanting sunbeams weren't considered much of a problem when we had plenty of oil, but now, when even respectable people are beginning to conserve energy, we're starting to demand our rights. We no longer want other people casting shadows on us, and many lawsuits are moving through the courts, saying thou shalt not shade thy neighbor's ox, nor his ass, nor anything that is thy neighbor's.

It doesn't require much imagination to foresee a time in the not-too-distant future when the inalienable right to collect sunlight anywhere on your property will become a matter of legal precedent.

Sun rights may mean death to many of the shady hedgerows of roadside America, and strange new shapes for buildings. And what confusion those laws will cause! Property rights go straight up, but sun rights must go up on a slant; not just a simple slant, either; the slant in Scranton is different from the slant in Atlanta. And all the slants change every hour—and every season!

What we'll need, to define our rights at any particular moment, will be property lines that sway overhead in the sun like fields of windblown grass. To describe them in a deed, we'll need the services of ten lawyers, a computer, and an astronomer.

Exaggeration? A little, perhaps, but it shows how easily selfishness could spoil even a miracle like sunlight. It shows how far-reaching the solar-changes in our energy policies will be. True, they won't be as far-reaching as a meltdown at a nuclear power plant would be, not by a long shot, but city planners are already predicting tilted skyscrapers and rotating buildings to take full advantage of the free, golden warmth. It's just possible that we'll develop the most beautiful, appropriate, and organic architecture this side of those master solar-structures, the trees.

The Window in the Winter

By Malcolm Wells

People often ask me how cars can consume so much energy. All I can say is: try pushing one sometime. If I pushed a well-lubricated car on a level roadway I suppose I could go 100 yards, but it would take me ten minutes and could cost me a heart attack. Pushing cars always renews my respect for the amount of energy in a thimbleful of gasoline. My respect for fuel oil would be just as deep, too, if I had to run the *half-million miles* each year it would take to heat my house by turning a treadmill.

Houses leak heat primarily through doors and windows. Even when double glazing is used, the heat-loss is impressive. In the Philadelphia area, the wintertime heat-loss through one square foot of double glass is roughly equivalent to the amount of energy you expend when you run 200 miles.

In other words, you could, by running 200 miles on a heat-generating treadmill, produce the amount of heat lost by your house through a small pane of double glass during a winter season. And since that season lasts only about 180 days in this climate, you could in that way free yourself of all heating-fuel problems by running about a mile a day—provided, of course, that your house had only one small window, that it was glazed with double glass, that no air could leak around it, and that the rest of your house had no heat-loss whatsoever.

Gives you a different slant on energy, doesn't it? Why don't you make a list of all your windows sometime? Then add up the heat-losses.

Just be sure to sit down first; the total will stun you.

Now for the good news: by a strange coincidence of geography and astronomy, the wintertime heat-loss through one square foot of double glass in the Philadelphia area is approximately equal to the amount of usable heat you can gather on one square foot of solar collector during the same period. No wonder so many manufacturers are rushing to get those things on the market. Imagine: no more fuel bills!

But whether you use a solar heater or an oil burner, the thing to remember is that all windows and doors should be tightly sealed against air-leakage, and that all glass areas should have better-than-usual heat-loss protection in the form of heavy drapes, or better yet, insulating shutters, which can cover glass whenever sunlight is not entering them, especially at night.

Our binge of glassiness has been expensive, to say the least. Our buildings leak their heat like sieves. That's why rising fuel costs, if not patriotism alone, demand that we do everything possible to conserve our dwindling supply of available, safe fuels until low-cost solar-heating products are in everyday use and we have been saved from the awful alternative of nuclear power.

Passive Solar Energy

By David Wright

Chances are most of the earth-covered solar houses you've seen published have been designed by this California architect. A former Peace Corps member, David Wright has written a book, Natural Solar Architecture, *which is a comprehensive primer on the principles of passive solar design.*

The widespread application of passive systems holds numerous benefits for the world community. Of principal interest is the potential of highly cost-effective space conditioning in areas where costly and power-dependent active systems are not feasible.

Passive solutions provide the opportunity to unplug from the use of energy compact high-temperature fuels for the low-temperature requirements of space conditioning and water heating. The development of simple, passive techniques for low-temperature commercial tasks, such as crop drying and distillation, will relieve some of the pressure on conventional fuels. The substitution of direct solar energy to replace high-temperature fuels, where possible, will reduce the net chain effects of the present inefficient production, transportation, and consumption of conventional fuels.

Take, for example, the popular irresponsible use of an electric clothes dryer versus the old-fashioned but efficient drip-dry solar clothes line, or the yet-to-be-seen solar clothes dryer. The more we can simplify our energy consumption, the more sophisticated will be our place in the hierarchy of nature. Direct-gain solutions, which minimize heat transfer and exchange, increase the overall efficiency of solar-energy use, and make better quantitative use of the solar potential.

Much of the population of the world lives in climatic zones that dictate the need for heating and/or cooling of structures for human comfort. A significant portion of people's time and energy is spent acquiring fuel for combustion in order to attain the degree of comfort necessary for social interaction. In various ways, we must all provide for light, space conditioning, food preparation, and water heating. The convention of gathering or purchasing wood, fossil fuels, dung, and other energy-storing forms is time consuming, whether done diretly by chopping and gathering, or indirectly by earning the exchange medium for trade. All present forms of combustion material either deplete nonrenewable resources, or place a greater burden on our renewable resources in this world of diminishing per-capita resources.

The agglomerate thermal- and air-pollution factor of all combustion processes is degrading our environment quickly and is as great if not a greater threat to the quality of life than resource depletion and overload. Which is to say, will we starve to death or go out in a cloud of smoke, as our present trend indicates?

The average peasant of South America, North Africa, the Middle East, or China spends a considerable portion of each day producing, obtaining, harvesting, transporting, and using fuel as a part of sustaining a subsistence living. This activity limits greatly his ability to obtain food, as well as limiting his involvement in activities other than survival. The subservience to survival may go a long way toward limiting the time and energy of the individual, thus hindering cultural, philosophical, and technical advancement, which are the ultimate hope for mankind's survival as a species.

If passive solar applications were implemented in over-populated or under-resourced areas of the world, a major burden of life's energies, as well as resource depletion and pollution, could be alleviated.

Sun tempering of structures can be very effective in many parts of the world. Just as climatic zones and solar potential vary, so do comfort standards. Our attitude toward comfort levels must be open and rationally viewed in terms of the dweller. For instance, recent evaluations of various passive systems have been conducted with an arbitrary acceptable comfort tolerance of 65° to 75°F. After having lived in an experimental direct-gain solar-tempered house for the last two years, I realize that a 20° temperature swing is not only tolerable, but perhaps beneficial. The standard 68° to 72°F comfort range is not only narrow for human beings, but could be unhealthy, in the long run. Our survival as a species is probably dependent on our continued ability to adapt (adaptability is a conditioned state). If we insist on a medium range in all extremes, we will cease to be able to function to the optimum in conditions other than normal. Can we always count on sustaining a narrow range of conditions? I contend that a 20° temperature variation in a radiant-conditioned space is not only more healthful but also certainly more cost-effective and practical on a low economic level. I'm not advocating that we all live in direct-gain or wide-temperature-swing habitats, only that we should be totally aware of human capabilities and limitations.

Much of the passive solar work here in the United States will soon affect major portions of the world. Let us concentrate not only on the most efficient and cost-effective methods of space conditioning, but also on water heating, cooking, lighting, desalinization, pumping, crop drying, and all of the other potential uses of solar energy, with an eye to what is really needed and with a broader view of reality than our technologically oriented standards have contrived.

The basic principles of sun-tempering structures are the most environmentally sound design criteria to be recognized in some time. Architectural evolution in residences and low-rise structures has not advanced much in the last hundred years, though the development of indoor plumbing and central heating were significant. However, spatial, structural, and functional advancement has been limited to systemization. The conventional home is still the basic balloon-frame structure. These buildings ignore or combat external design factors. The popular tradition of constructing Cape Cod or colonial imitations, from New England to Texas to California, has not changed appreciably with the contemporary renditions of versions. Even stud-frame-imitation adobes proliferate in the developments here in the Southwest.

The time for re-evaluating basic design criteria has come. The influence of careful planning to accommodate nature and varied climates is a major step forward for architecture. The overall application of climatic design and placement of structures will change the architecturescape of this country in a relatively short time. A new ethic of design suited to specific microclimates and regions, rather than imitative styles and solutions contrived to *combat* the local environment, will be a fresh new approach to shelter.

Primary to climatic design is the idea of designing structures that interact most favorably with the specific natural factors of a particular site. Many new variables will influence the designer's decision-making processes. In the past, a simplistic attitude of calculating the heat loss and designing a closed compensating system to satisfy temperature and humidity standards was adequate. Probably, view factors and automobile accommodation were the major site considerations beyond function, economics, and structural design. In the future, before the inside of a building is designed, a multitude of external factors will help to establish the form, siting, and structure of the building. Seasonal wind, rain, fog, snow, temperature, solar, air, cloud cover, and other factors will be ascertained, both quantitatively and qualitatively. The surrounding vegetation, soils, terrain, drainage, traffic patterns, noise factors, and solar occlusion will be studied and evaluated. By systemically following these external influences, much about the physical visual form of any structure or complex will be given to the architect, designer, and engineer before planning the interior functional form and plan. This is in many ways a reversal of the traditional design-development procedure.

The solar esthetic or climatic design approach, if logically followed, will produce a new kind of architecture. We will not only have buildings that imitate nature in her innately perfect design ability, but also will have buildings that should be far more beautiful than the contrived forms of the past. Structures that are totally harmonious with their surroundings will flow into the countryside or cityscape as though they belong together. They will complement one another and will fit into a natural master plan. This is a far cry from the hodge-podge hit-or-miss planning methods of the past. Of course, this attitude will create endless variety, the only redundancy being dictated by the world of nature; a visual tranquility and suitability should prevail.

All structures should be designed to function primarily passively. Each should be analyzed to operate at various seasons of the climatic year with the minimum amount of conventional or active solar-conditioning. After all of the insulation, shading, ventilation, solar-absorption, storage, humidification, and insulation factors have been optimized, then the requirements of augmenting these by conventional or active solar mechanical systems should be established and integrated into the design. With this environmentally conscious design approach, we can some day hope to modify our dependence on commercial energy. Further, the cost of equipment, maintenance, and operation of structures will be greatly reduced—this, of course, is probably the best argument the designer can offer to the developer/builder to justify environmental design.

The climatologically designed, passive, low-energy structure will be visually logical—integrated with its environment and readily viewed by even the untrained eye as an object at peace with, and a part of, the whole. As people become more aware of the variations, benefits, and logic of this approach to architecture, their appreciation will grow, and a more refined basis for esthetic qualification should be realized. The esthetic value aspect of human judgments has much room for improvement. That is not to say that what has been traditionally practiced or valued is not valid—quite the opposite is true. It is only by trial and error and experimentation that we experience knowledge. The architecture of the past, from the Anasazi Indian cliff pueblos to the high-rise skyscrapers, is a part of our experience, and we must recognize and retain the best working aspects of it. It is now time to evolve beyond our traditional concepts to a more enlightened architecture. Progress will be slow—education and examples are the keys—the seeds are just now being sown and we must commence with a new ethic.

Technological Versus Natural Destruction

By Robert Finch

*My New York-centeredness was showing badly when I discovered the
writings of Robert Finch. I'd just moved from New Jersey to Cape Cod and
was surprised to find a writer of his stature doing a column for the local
paper. But why not? His view may be universal but his facts are usually
local—and* The Cape Codder *is an impressive paper. Bob's now at work
on a book about man vs. nature for W. W. Norton Company.*

Recently, the earth's oceans experienced a seeming epidemic of oil
spills, beginning with the grounding of the tanker *Argo Merchant* on the
Nantucket Shoals and the subsequent loss of 7,600,000 gallons of
crude-oil cargo into the North Atlantic. Nearly each week after that, a
fresh spill was reported on our front pages—in Chesapeake Bay, Long
Beach, Hawaii, the Bahamas—until after a while they seemed as
routine as the Sunday crossword puzzle. They became something of a
fad, complete with jokes about "oysters Rockefeller" and refining seawa-
ter, perhaps because many of us needed to find something to laugh
about during the periods of extreme cold that gripped much of the
nation.

　　With the coming of spring, the oil spills gradually "went away,"
just as the initial energy crisis "went away" in 1974 (though those of us
who live near the shore do not expect we have seen the last of clean-up
crews and befouled seabirds). Other news has supplanted them in the

headlines, and scientific attention is now focused, as it should be, on such areas as the monitoring of ocean sediments and the study of suspected changes in the behavior of lobsters and other petroleum-affected species.

However, other questions raised by the spills have received little attention. One was dramatized by a newspaper photo, which, for me, remains the most striking image of the entire, oily winter. It showed enormous, black plumes of smoke rising up from the frozen white surface of Buzzards Bay, at the south entrance to the Cape Cod Canal, as the Coast Guard attempted to burn off some of the 100,000 gallons of No. 2 heating oil that leaked from the grounded barge *Bouchard No. 85*. That picture captured the winter's irony: spilled oil burning on the ice while factories and schools were shut down in Ohio and Pennsylvania for lack of fuel. But it also posed, almost symbolically, the difficulty in distinguishing between technological and natural disaster.

The problem is not just academic. The federal government makes just such fine distinctions when it hands out disaster relief money, as we saw soon after the *Argo Merchant* wreck. Governor Dukakis's initial request to have southeastern Massachusetts declared a "natural disaster area" was turned down because the spill was "man-made."

On the other hand, the city of Buffalo, buried in snow, quickly received the "natural disaster" designation. But it seems to me that this disaster was as much the result of human error (of underestimating the chances and the punch of such a winter, and failing to prepare for it) as any delinquency on the part of the *Argo Merchant's* captain. After all, both disasters represented unforseen conjunctions of human enterprise and natural processes. A miscalculation in navigation in July would probably not have resulted in any significant spill, and an unusually severe snowstorm in upper Labrador would hardly have rated any headlines.

Yet, though governments and the media make such distinctions, it is often not so easy to explain our different reactions to the effects of human and natural destruction. Why are we more distressed to hear of thousands of fur-seal pups slaughtered and skinned by pelt hunters in Newfoundland than we are by the thousands of pups slowly dying on the rocks from hookworm infestation? And why are we more likely to see pictures of one in our papers than the other?

The death of 15,000 sea herring at the entrance to the Pilgrim Station Atomic Energy Plant in Plymouth, Mass., received wide attention in the press, and rightly so. Yet our interest in the stranding of millions of long-finned squid on the beaches of Cape Cod Bay lapsed once chemical pollution was ruled out as a cause.

Who says nature is kind?

Mortality is the rule in nature, and we tend to accept it, perhaps unconsciously, with indifference or philosophic acquiescence—as long as it doesn't touch our own interests too directly. On the other hand, human-caused destruction in nature tends to raise our hackles.

The juxtapositions frequently seem designed to confound us, especially in a place like Cape Cod, which is so vulnerable to both human and natural forces. Our tern colonies, which find their nesting sites on low, exposed beaches, are annually threatened by motorboats, beach buggies, and foot traffic. Yet hundreds of these tern chicks are also baked to death each year in the summer sun, or are swallowed in the rising gorge of a summer storm, or are decimated by a few great-horned owls. The Cape Cod Museum of Natural History treated some two dozen murres, auks, and other pelagic birds that had been stained by assorted oil spills. And just the other week, walking an isolated strand of barrier beach, I found the carcass of a brant that was emaciated, oil-stained, and viciously decapitated. To what was I to chalk up this death: ice, oil, dogs? Where, after all, does man leave off and nature begin?

Some, who would use the undeniable magnitude of natural destruction to explain or justify our own, suggest, for instance, that depleted fish stocks are as much the result of natural population cycles as of overfishing. They imply that we have just as much right as nature to destroy, and what is all the bellyaching about anyway? Who says nature is kind?

Others, frightened by the apparently burgeoning effects of human technology and numbers on the world, seek to quarantine us from the environment like an unnatural infection. Yet we cannot, even if we would, live like a flock of sanderlings, running perpetually along the surf-edge of the world, nimbly picking up our nourishment without getting our feet wet. Even if there were a way to quarantine ourselves, we could not do it. People need and demand participation in nature. Without it we atrophy.

If we concede that it is frequently difficult, if not impossible, to distinguish between natural and human destruction, is there any basis on which we can make a useful distinction? I find a possible answer in the current trend among environmental groups to move away from so-called "focal-point" conservation. These environmentalists are less interested in saving specific endangered animals or sites—whooping cranes, red wolves, Franconia Notch—than in a broad effort to protect and conserve entire systems of natural processes. By emphasizing processes rather than places or populations, we begin to recognize the source of all natural health and abundance, without which no bird or forest would exist. We are learning that, in the long run, it is more important to protect the larger, creative processes of nature—ocean currents, sand transport, marsh development—than, say, a particular stretch of beach or flock of birds, which may be only transitory and purposefully vulnerable parts of the whole. Technological and natural destruction cannot be distinguished on the basis of their immediate, observable effects. Even the term "destructive" is largely subjective: one man's beach is another man's eroding shore. The critical difference is that any natural force is linked to a myriad of other forces around it—physically, intimately, and immediately linked. Every natural process is part of an unbroken web or matrix of connected events, accountable to and cooperative with each other. This is true of an avalanche roaring down a mountainside (perhaps wiping out an alpine village with it), or of a bluefish chopping its way through a school of mackerel (perhaps taking a fisherman's artificial lure in its blind greed). "Destruction" in nature—whether of a beach, a school of fish, or even an entire ecosystem—is characteristically a necessary part of a larger, balanced system of natural processes.

Not so with human disruptions in a highly technological society. What distinguishes our destruction from nature's is its characteristic separateness from what it destroys, its lack of interaction, dependence, or accountability toward what it affects. What, after all, steered the *Argo Merchant* onto the treacherous, fog-shrouded shoals? What pushed the beach buggy onto the fragile dune and into the vulnerable tern colonies? What brought the nuclear power plant to the shores of Plymouth?

The answer is human abstractions: abstractions in the form of international finance, flags of convenience, corporate planning, economic trends, and recreational fads. Add to these our singular ability to transfer large amounts of energy from one place to another without a physical accountability, a balancing of forces, at either end.

This is why, I think, a disaster like the *Argo Merchant* spill holds such terror for those of us who live near it and learn something of its nature. It is not just the hundreds or perhaps thousands of sea-birds that died from pneumonia or internal poisoning, nor the possible eventual harm to shellfish and fish populations, nor the loss of revenue to fishermen or to motel owners if the sticky stuff ever reached our beaches.

The fear is that, if we cannot contain such disasters, what assurance do we have that nature will? Hurricanes spawn, grow, and blow themselves out. Floods subside, earthquakes redistribute geologic forces, and fires consume themselves. But with oil spills, there is no similar point at which we can say, "It is over; we have survived." They are new disruptions, with no history or life-cycle we can count on, with no race-memoried assurance of whether, when, and how they might settle themselves out. Such man-made disasters have no life, and therefore no death, of their own, no beginning in nature and so no end. They are deadly passive, borne along by the very natural processes they contaminate, infiltrating the structures of life. And so, although the *Argo Merchant* has slipped beneath the waters and out of our headlines, its effects and those of the progeny it spawned bode to spread out endlessly, composing sad, tedious tales of lethality.

Nuclear and Solar Economics: A Paradigm Shift Is in Progress

By Hazel Henderson

"This 'econo-clast,' " as Alvin Toffler of Future Shock *calls her, "seems to be made up of a good chunk of small-is-beautiful advocate E. F. Schumacher, plus bits of environmentalist Barry Commoner, consumerist Ralph Nader, and visionary Buckminster Fuller." Writer, lecturer, futurist Hazel Henderson, former co-director of the Princeton Center for Alternative Futures, is the author of* Creating Alternative Futures *(New York: Berkley, 1978).*

The issues surrounding the use of nuclear power or the use of solar power—and the implications of the choices between the two—are symbolic of the sharpest differences between the two directions lying before us, toward greater and greater capital, energy, and materials intensity, or toward greater labor intensity. The current direction, which was historically sensible, overshot the mark. Saving labor by making a system more capital-intensive is reasonable when you have very cheap resources and not much of a problem in putting those resources at the disposal of workers for increasing individual productivity. But this system has now collided with resource scarcities.

The entire economy, the whole configuration of factories and cities and suburbs, laid out in concrete, gives the system tremendous momentum in the existing direction—and exploitation of nuclear energy is a last, baroque elaboration of that old direction, no longer sustainable. Solar, of course, is the key metaphor for the way we have to go. So the situation is polarized around these two types of technology.

I see all kinds of external forces pushing the economic system into a desirable new state. Whether or not we manage to correct some of the major subsidy programs built into the system—those that keep pushing it toward greater capital-intensity—whether or not we work out an equitable way for energy prices to rise without hurting too many poor people, there are many ways the economy is being driven toward the new state that has nothing to do with human beings and our attempts at policy-making. Availability of energy, of course, is the main driver. The system's own pathway of accommodation is expressed as inflation. Barring any conscious policy, inflation will drift up and quietly settle down into a more stable sort of economy and a less centralized pattern. This course will be very difficult for some groups, and there will be a tremendous amount of unnecessary pain in simply allowing the system to do its thing. But, if the pattern of events can be explained in ways that people can understand, we might not have to go into such catastrophic sorts of over-shoot as continuing to hype the old industrial sectors, subsidizing declining corporate dinosaurs, and trying to consume our way back to prosperity by stimulating demand for cars and TV sets and other energy-intensive goods in the hope that this will trickle down and create jobs.

There are two issues that I would try to illuminate in very broad terms. First is the matter of subsidy. We have subsidized every other form of energy technology except solar and thereby have gotten into a bind: either we must subsidize solar equivalently, if we are going to be fair, or we must reduce the subsidies on other technologies to allow solar to compete. If we are to be efficient, we eliminate subsidies. "An Analysis of Federal Incentives Used to Stimulate Energy Production," just completed by the Battelle Institute, is an attempt to determine the extent of these subsidies. It turns out that, even though subsidies to nuclear power have been enormous, they have not been as great as the subsidies to oil.

Oil subsidies include depletion allowances and all of the special long-standing tax legislation. Companies are dependent on these subsidies now. The subsidies have built up an enormous client system that is very, very efficient at lobbying to maintain its advantage. Politically, it is very difficult to remove the subsidies.

We are therefore in the very socially inefficient business of having to subsidize new energy technologies equally. The distorted market price of old technologies makes them inefficiently cheap from a societal point of view, so unless the new ones have sufficient subsidy, they cannot produce the innovation society requires, particularly when—as is the case with solar—they frequently will have to interact and compete with the established energy industry.

We can pursue the problem on either or both of two fronts and do the best we can to illuminate the situation. We must work as hard as we can politically to phase out those old subsidies gradually, while trying to reduce the pain to innocent individuals—and at the same time try to subsidize the newer technologies rather quickly so they can get a foothold. The new and the old energy systems are fighting it out in Washington now, but the leadership is coming from the states. I have begun to see how a state like California, with 10 percent of the population of the entire country, can force the situation. The steps taken with California's 55-percent consumer tax credit for solar, and other steps the state is taking unilaterally, are going to help shift the entire country into the new pattern. If you really make that enormous market fair to solar and renewable energy, it becomes a test bed in which companies can commercialize renewable energy sources and develop economic strength to lobby in Washington. Because California has so much incident solar-energy income every day, and because its leading industry is agriculture—which means that the people are much closer to the real biological efficiency of the system—I can see California becoming the state that will lead us into a renewable-resource economy, the pilot project for how that transition is going to work.

At the same time, nuclear power in California, with the defeat of the Sun Desert nuclear plant, is really dead. Any new electrical capacity in California is going to be coal-fired with the best state-of-the-art pollution controls, to tide the state over as it builds toward a renewable-resource energy user.

The second issue is this: The energy debate now going on is too narrowly defined. Many energy enthusiasts hold that we just have to load the real costs into the price of petroleum and the other old energies to shift the system toward greater labor-intensity. That is certainly one policy option, but in my view the solution is more complicated than that. Central is the fact that the energy sector is part of the whole economy, and the paradigm, the pattern by which we understand what's happening, has broken down for our whole system of economic mapping, which is why macro-economic management is failing.

It is useful to think of a paradigm as a particular pair of spectacles through which you view the world; when you take that pair off and put on another pair of paradigm spectacles, you see phenomena to which you paid no attention before, different features of the whole system. The dominant paradigm in economics, the equilibrium market system in which supply and demand balance, now has so many exceptions to the rule that it's positively embarrassing. For instance, there is the set of exceptions called externalities, those costs visited on innocent third parties—air or water pollution, the increasing size and scale of our cities, medical problems, systemic disruptions—that in the market-model world are considered to be side effects, slight aberrations not to be worried about. But you might now say that we have arrived at a point where the economy is better described as one vast set of social costs, a mirror image of the GNP, perhaps the only part of the GNP that's growing. The problem is that we add them to the GNP as if they were real products, and so the GNP keeps merrily going up.

In the dominant market model, another example of a paradigm coming apart is that technology is considered to be an external variable. Yet we have now reached the point where the very size and scale of technologies such as nuclear begin, systematically, to destroy the conditions of free markets.

The first of two basic conditions under which a free market can allocate resources efficiently, described by Adam Smith, is that buyers and sellers meet each other in the marketplace with equal power and equal information. The second condition relates to externalities: that no costs of the transaction be visited on innocent bystanders or externalized to the social system. This market system isn't functioning well any more. More and more centralized and capital-intensive technologies require the kind of centralized economic and political power that destroys these free-market conditions. In looking at the problem through the old paradigm spectacles, economists do not understand that, basically, the United States economy has become a vast system of income and wealth transfers instead of a free market, that we have consciously legislated this system of allocating resources, though we pretend that it's God doing it via the "free" market, the "invisible hand"—now the most pernicious idea clouding the thinking of almost everybody who thinks about the American economy. Markets do not derive from God, or, as Adam Smith thought, from a propensity in human nature to barter. Even though humans have always bartered in local marketplaces, the idea of setting up a national *system* of resource allocation relying *only* on markets was a very new idea first tried in Britain about 300 years ago at the start of the Industrial Revolution. With this tremendous misunderstanding, we now run around legislating markets, putting on tax incentives and penalties, and incurring all the external costs, which we then refuse to add into the final price. And we look at energy policy as if letting costs rise to the cost of replacing the supplies, or to the world price, will in itself encourage the innovation of renewable resources, discourage the use of old energy forms, and shift the entire economy into a pattern of great labor intensity and less energy and capital intensity.

It may be *necessary* to increase the price of energy, but it will not be sufficient. Other major forces in the society are shifting us in the direction of greater capital intensity; without considering those forces, we won't achieve our purpose. The biggest economic force to consider is the investment tax credit. Alone, it can go on skewing the economy and smothering the effect of rising energy prices. A study quoted by Senator Kennedy of the Joint Economic Committee showed that the 1,000 largest companies (according to *Fortune's* 500 standards) used 80 percent of the total tax credit and 50 percent of all industrial process energy, and only created 75,000 new jobs over seven years. In the same period, the country's 6 million small businesses, using far, far less process energy, created 9 million new jobs.

The tax credit was originally justified as a means to create jobs. We find that with the largest and most capital-intensive companies, investment just as often disemploys people, through automation, or through moves abroad. Over the past five years, I have worked to expose the tax credit as the enormously destructive boondoggle it is. As a member of Jimmy Carter's Campaign Economic Task Force, I suggested that, first, we should try to make the tax code neutral between capital and labor, with tax credits for creating employment, not for investment per se. That idea was put into one of Carter's tax bills and was actually reported out of the House Ways and Means Committee in the first year of the Carter Administration. It was knocked out in the Senate Finance Committee.

Here is a perfect example of a paradigm problem, which might be described as a war of paradigms about how the economy works: between the older paradigm of the "golden goose" and the newer paradigm of the "milking cow." The latter holds that in the stage of industrialism we have reached, the federal and state governments have become the private sector's milking cow, and that most companies prosper or do not prosper according to whether they can rig the system of tax subsidies and the markets to favor themselves. The older golden goose paradigm (reasonably accurate in the early Industrial Revolution), says that companies in the private sector create most of the jobs and all of the wealth, and governments should only maintain a good business climate, keep its regulatory hands off, and let business do its thing in a "free" market. Paradoxically, this denies the social costs incurred by business activities, while business lobbies for tax subsidies as well! The golden goose view won in the Finance Committee. Now, as Peter Harnick, Byron Kennard, and I recognized when we created the Environmentalists for Full Employment coalition, the only way to break that political log jam is to organize a constituency around the new paradigm.

To make matters worse, we have created federal-level policy that is going to push the system into greater use of energy and capital, and cause even greater unemployment and inflation. I am referring to the new Social Security legislation, which raised Social Security taxes disastrously. When you raise Social Security, you put another tax on employment. Employers are automating more, because it is becoming so expensive to hire people. The energy system, it is evident, is embedded in the much larger system of transfers that make up the whole economy.

The entire economy, then, must be shifted toward a system that combines more people with less capital, energy, and material. How can this be done? Howard Odum, at the University of Florida, states the problem well: the energy flowing through any system maintains its structure. The moment you begin to withdraw energy, there is a spontaneous devolution of the structure to a level appropriate with the new, lesser, energy flow. You could look at the decentralization already going on in cities, and in the economy, as a demonstration. Neighborhood economic development becomes more efficient—as does any flatter capital structure that has to service fewer stockholders, smaller office buildings, and no company jets. It made no sense to bake cookies on one side of the country and sell them on the other.

A number of industries, operating under the old paradigm, are making counter-productive decisions, because the tax system still drives them in the wrong direction. Insurance companies, for instance, are beginning to be sensible enough to bail out, not to try to insure the unnecessarily high-risk energy technologies that are building up, whether it is nuclear power, where they are protected from risk by the Price-Anderson Act, or ultralarge oil tankers, for which 70 percent of the operating cost is insurance, or LNG transports, which incur fire risks for which no fire-fighting technology exists. Operators and promoters of such high-risk technologies often set up captive insurance companies, by which a subsidiary insures them against technically uninsurable risks. Then such operators request tax-exempt status for these captive companies from Congress, thereby socializing the cost, adding it to our economic "externalities."

The report done by Dr. Norman Rasmussen for the Nuclear Regulatory Commission (WASH-1400), contained an enormous number of errors in risk calculations for nuclear accidents, but there was an underlying error in the difference between an equilibrium system—a cybernetic system in a steady state—and a morphogenetic system, one moving toward a new state, governed by positive feedback loops. In calculating probabilities for insurance purposes, equilibrium conditions are assumed, leading to the assumption that the greater the accident, the less likely its occurrence. In real life, we are now dealing with a disequilibrium economy that piles risk on risk. You could just as easily assume the greater the accident, the greater the probability of its occurrence. We don't know how to model probability under these conditions; the underlying logic of insurance breaks down, and we merely socialize more and more risks. This in itself would be a good rationale for being very conservative about energy technologies—underwriting conservation and renewable energy where the risks are in the knowable range, rather than technologies where the multiplication of risks is totally unmodelable. If the insurance companies won't insure it, or if its risks can't be socialized, it cannot be built—and there is no further justification of ever more incalculable risks when there are many safer alternatives.

There is real and substantive discussion going on about methods for shifting the economy toward a renewable-resource pattern, and there is a general understanding that this society—and any other industrial society—is too energy-intensive, too materials-intensive, too capital-intensive. To use an earthquake analogy: the more pressure builds up, and the more wrong decisions are made more quickly, the faster the system will shift toward its new state. The process can hurt many people. The Princeton Center for Alternative Futures has just published *The Inevitability of Petroleum Rationing in the United States*, by Carter Henderson (available from the Center, 58 Hodge Rd., Princeton, N. J. 08540; $4.25), in an attempt to chart one path. This proposes that the only equitable way to share our petroleum is through portions allotted to every American over the age of eighteen, for use or for sale as he/she sees fit. Franklin D. Roosevelt, in a discussion of rationing before the Second World War, said that Americans are too democratic to stand for rationing by price alone, that rationing by commodity was the only equitable way to do it in a democracy. Such rationing would be a beginning.

Boom!

By Malcolm Wells

Washington, D.C.—(UPA)—1980. According to high administration sources, the President's move to demonstrate nuclear power-plant safety by offering to build a full-size atomic reactor inside the dome of the Capitol has left other world leaders years behind in the energy race.

The proposal is expected to give a long-delayed boost to the nuclear power movement, and has, at least temporarily, stunned into silence congressional and environmental critics alike.

By 1981, the thousand-megawatt reactor inside the Capitol will contain the radioactive inventory of over a hundred Hiroshimas. When asked by reporters about the frightening possibilities of disastrous accidents—explosion, sabotage, vessel rupture, meltdown, earthquake, or the leakage of radioactive materials—a White House spokesman pooh-poohed such worries and called for a greater measure of faith in the judgment of our political leaders. His appeal was followed by a sharp jump in the stock prices of the energy companies involved in nuclear power.

Not to be outdone by the President, the governors of more than 40 states have by now declared intentions to build similar reactors inside their own capitol buildings. "Nuclear reactors have always before been built far away from population centers because accidents out there would kill only a few farmers and cattle," said one governor, "but now all that has changed." Later, his press secretary issued a clarification, stating that the governor had nothing but the highest respect for farmers and livestock, and that nuclear power was now 100-percent safe, just like air travel.

Most recent of those boarding the nuclear bandwagon is the mayor of a large suburb, who announced late today at a hastily called press conference that he planned to go even further than the President and the governors by accepting bids for a nuclear power plant to be built in his very own basement. Noting that "in this age of infallible computers, 'lemons' have become things of the past," the feisty mayor closed his statement by calling his decision "one small step for me; one giant leap for mankind."

As a further show of confidence, he encouraged trucks carrying supertoxic plutonium wastes to pass freely through his town. "Let other towns ban the transport of nuclear wastes; you won't be banned from ours. Come on down," he said, "We don't believe all that danger-talk." *(The mayor was referring to recent scientific evidence that as little as one-millionth part of an ounce of plutonium can cause an incurable type of lung cancer. —Ed.)*

Industry observers, noting the public's placid acceptance of these developments, predict that fission reactors will soon be used in the most densely populated parts of all major cities, making the energy crisis a thing of the past, and paving the way for a nuclear boom beyond our wildest dreams.

Heating Up

By Malcolm Wells

My son Sam suggested that I write about things nobody else knows. When I challenged him to name one thing he knew that nobody else knew, he didn't hesitate a bit. "My shoe size," he said, and I had to admit he'd stumped me.

But I like to write about the kinds of things everybody knows and nobody thinks about; familiar things like those free heaters and air conditioners they're giving away down in Miami this week. Forgot about them, didn't you? You forgot that each of us—in Miami or Philadelphia or wherever—arrives on this planet fitted with a completely automatic body heater that can run on vegetables for 80 years. Your heater's running right now, in fact. The point is, are you using it properly?

When Ernest Shackleton and his men were shipwrecked in Antarctica, they spent weeks in an open boat, soaked to the skin in zero weather, warmed only by the heating systems in their bodies. The systems worked so well the men felt uncomfortably warm whenever the temperature got as high as freezing! That's what you call learning the hard way, but what a lesson they left us!

When I walk to work in the wintertime I start out buttoned to the neck, scarved and gloved. Halfway to the office, the scarf's off, the gloves are pocketed, and my coat is open. An hour of walking gets my heater running.

But a few hours later, sitting at my desk, I begin to feel cold again, even in this stuffy 65-degree temperature. Why? Because I've let my heater run down, that's why, and I'm too lazy to walk around the block to get it running again. Instead, I too often sneak over to the thermostat and turn it up when no one's looking. (Conservationist!)

Most modern buildings are so poorly insulated that the little body-warmth we produce seems insignificant. We're forced to depend on artificial heat and big fuel. We play right into the hands of the gas, oil, and electric companies. They don't like us to think about body-furnaces. Can't sell fuel that way. No, but they know as well as you do that in properly-designed, properly-insulated spaces we might not need much fuel at all.

And, in the summer, with a bit of shade, some ventilation, and the right insulations, our body-cooling systems can keep us fairly comfortable, which is probably about as comfortable as we're supposed to be: fairly. We can't seem to afford total comfort. Total comfort is too expensive, not only in money, but in fuel-smoke, nuclear radiation, oil spills, and heart attacks.

So why don't we all wake up and use the gifts we were given? I don't know. You tell me. I don't have enough guts to really face the question.

Blood, Sweat, and Tears. And Spit.

By Malcolm Wells

I'm really happy.

After what seemed like hours, the dentist has finally pushed the folding drill-arm all the way back. That means the rest of this appointment will be pure pleasure; nothing but cotton wads, fillings, and squirts of tasty mouth-wash. But he seems to think we can now carry on a conversation. He asks if I'm growing my own vegetables this year, and all I can say is "uh-uh". (I'ave one 'ose 'uction'ings in my mou'h.)

He takes my answer as a sign of disinterest and starts doing something on the counter behind me. There's no further talk, and that's all right, because I've just discovered how to play tunes on this spit-siphon. I do a few bars of "Star Spangled Banner," and they come out beautifully until the dentist turns back and gives me a funny look, so I pretend I'm just rearranging my mouth into a more comfortable position.

As the sucking sounds rise and fall I think about this saliva I'm losing. I'm wondering where it all comes from. Sifting back through my sketchy knowledge of anatomy I can't remember ever hearing of a water-circulation system inside the human body. The only such system I know of is the one for blood. So the saliva glands must take the water they need directly from the arteries. But wait a minute; if that's true, then teardrops and sweat must be made in the same way.

I calculate how much sweat I can work up on a hot day. Quarts! If it all comes from my blood it's a wonder my corpuscles don't find themselves rolling along dry, dusty veins all summer.

The dentist is pulling the wads and things out of my mouth. Now I can ask him about it . . .

Well, I won't drag you through this long conservation on spit. And sweat. And tears. It goes on for 15 minutes, and I don't know whether I'm more impressed by his knowledge or by what he's telling me about the fantastic recycling systems we all carry around in us.

Saliva enters the mouth as an utterly clean liquid but it is immediately polluted by all the bacteria and other goodies that live there. Every time I swallow this germy soup—as I do, unconsciously, all day long—it gets sent to my kidneys, where all the impurities are removed before the water is sent back into the blood stream again.

And that's just one of the miraculous recycling systems in my body. They're built to last as long as 100 years, they can repair many of their own defects, and this whole crazy engine runs perfectly well on nothing but air, water, and food. There are even control organs that keep the blood from getting too thick—or too thin. You'd think drinking a few quarts of water would dilute blood to the consistency of tea, but it doesn't.

We must be in good hands.

It's strange to swish this familiar liquid around in my mouth and realize that just a few minutes ago it was riding through my feet and my brain as part of my blood, and that when I swallow my saliva it will be on its way back into the endless circuit again. It's like a mini-version of the earth's own water-circulating system: rain to earth, to mud, to root, to branch, to leaf, to cloud, and back to rain again. It confirms my belief that if we are ever to get rid of our poisonous, wasteful ways, we'll be wise to take closer looks at these already perfected miracles.

Now my dental appointment is over, and I feel really good. Alive. Part of the feeling, I guess, comes from the knowledge that I'll be away from the drill for another six months, but most of it, right now, is from my intense awareness that this fleshy bag of me is a rare gift indeed. The dentist calls the human body "an awesome piece of equipment," and that's one of the nicest things anyone has ever said about me.

We'll Have the 5-Day Forecast After This Brief Message . . .

By Malcolm Wells

. . . *aspirin is good. Beer is good. Airlines are good. And cars are good. Diet drinks are good. Deodorants are good. And gasoline is good. Bad things are things like gray hair and indigestion. Good things are things like Living Bras, Listerine, VapoRub, and Maybelline.*

WE RETURN NOW TO YOUR LOCAL STATION FOR
A LOOK AT THE WEATHER.

Thunderstorms are bad. Blizzards are bad. Hurricanes are very bad. And tornadoes; they're awful! Bad weather includes sleet, hail, rain, drizzle, humidity, and heat. Also cold weather, chilly weather, and those rare bright days that make you squint.

From what I can deduce, good weather, in the eyes of our TV weathermen, consists solely of those airless days during which greasy sunlight filters down through all the smoke. "Good weather" is barbecue weather, football weather, suburban weather, American weather. Don't knock it; we like it like it is.

Early last spring, in New Jersey where I live, there was a record dry spell. For almost a month there wasn't a trace of rain. Toward the end of the month, people began to talk knowingly about the drought, about how bad it was for the crops (meaning their lawns). And when the rains finally came everyone said, "We sure needed this."

But it kept on raining. Three days. Four days. And we all went sour again. At first there were good-natured storm jokes ("Wet enough for you, Fred?") mixed with knowing recitals of last night's rainfall statistics. But as the weather got wetter the jokes dried up and gave way to our true feelings. "Isn't this lousy?" "What a rotten weekend!" "Whew! It's really awful outside." All this in spite of brimming reservoirs, lush green lawns, and that fresh, moist smell everywhere. Nature was shouting her delight at us, and we were calling it lousy just because the weathermen—the authorities—had taught us to.

Well, I say those authorities ought to be deported. I can think of few more insidious disservices to mankind in these times than the preaching of hatred for the miraculous sources of all earthly life; air, water, and sunlight. It's insane. Not only national policy-makers and voters but also millions of innocent little kids and their teachers are reminded two or three times a day that rain, wind, fog, sleet, snow, cold, and every other meteorological phenomenon except smoky blah weather are things to hate. It affects everything we do—how we think, where we go, and how we feel about life and nature.

And it's getting worse. If false values continue to be preached through a *second* TV generation, those bland nitwits—the weathermen—who in utter innocence originated these values will have made major contributions to the final destruction of the land.

Now just imagine, on the other hand; what might happen if philosophers or poets were to replace the weathermen! After all, how many of us need to know the exact wind speed or the second decimal on the barometer reading? Wouldn't it be nice to hear about glorious sunsets for a change? Or to see the new weatherman come puffing into the studio, red-cheeked on a wintry day, to describe the mile walk, through three-foot drifts, that took him an hour to complete? Or suppose he said the kind of things Henry Thoreau used to say: "There was never yet such a storm but it was Aeolian music to a healthy and innocent ear. The gentle rain which waters my beans and keeps me in the house today is not drear and melancholy but good for me, too. If it should continue for so long as to cause the seeds to rot in the ground and destroy the potatoes in the lowlands, it would still be good for the grass on the uplands, and, being good for the grass, it would be good for me. Some of my pleasantest hours were during the long rainstorms in the spring or fall which confined me to the house for the afternoon as well as the forenoon, soothed by their ceaseless roar and pelting; when an early twilight ushered in a long evening in which many thoughts had time to take root and unfold themselves."

We all know by now that snow can be miserably cold and that hot, humid weather can be uncomfortable. But why make them worse? Are there so few real troubles to concern us that we've got to try topping each other by comparing thermometer readings? My greatest hot-weather discomfort has come, invariably, from listening to the weatherman. When I am sane enough not to listen, the mercury seems to drop by ten degrees. The natural air-conditioning systems in our bodies (remember?: perspiration, aspiration, circulation) may not be able to hold us in exact 70° comfort, but they're fully automatic, they require no batteries, and they often keep themselves in working order for as long as 70 or 80 years. The same systems keep us warm in winter too. So why is there all this terror over each dip and rise in what is perfectly normal summer and winter weather at each particular latitude?

I was almost 40 years old before I ever walked barefoot in the snow. From all the years of snow! disaster! warnings on television I half expected to see my toes drop off the first time I ventured out. But I found to my surprise and delight that the experience was great—for five or ten minutes, that is. It was like walking on the crunchiest, softest, and coldest kind of sand. I didn't stay out too long that first time, but I stayed out long enough to realize that I was walking on millions of crystals, any one of which could have put a jeweler to shame. And I discovered what poor traction my boots had been providing all those years. With five toes to help it grip and dig in, each foot really grabbed the snow. Maybe winter footwear should look more like gloves than boots; I don't know. I do know that each time I go barefoot in the snow, the crazy footprints alone justify the whole adventure, and I am made aware that I'm knee-deep in a miracle again.

We're no longer forced to spend weeks at a time battling the elements. But it doesn't follow that we should spend *no* time experiencing them. Our waterproof skin, our lungs, our hair—even the shape of our noses—were all meant to know a variety of weathers. We were meant to be attuned to the life forces that swirl around us. Instead we seem to be attuned only to CBS and NBC. As a result we consider threats like insecticides, auto exhaust, and sewage-filled rivers only as threats to ourselves and not, as they truly are, as threats to all the animals and plants that have ancient and equal rights to eat and breathe and drink here too. It's the sickness of man that we continue to see every-thing on the TV weatherman's kind of how-will-this-affect-me scale. As long as we ban DDT only to save our own species we'll get nowhere; we'll simply find other ways to poison wild creatures—and wind up, of course, poisoning ourselves again.

There is no hope except through whole new motives and concerns, through things like seeing weather as an endless pageant of beautiful miracles. ("Look, Ma; it's starting to rain! Can I go out? Please? Can I?") A very accurate gauge of our ecological health is the attitude of the TV weatherman. The day we hear him trying to describe the exact colors of a stormy sky or speculating as to how good those raindrops must feel to drought-shriveled leaves is the day we'll start to turn around and wade out of this mess.

Winter

By Malcolm Wells

As if Christmas expenses, on top of inflation and recession and fuel shortages, weren't enough to worry us, now we have winter. In a few days, the weatherman's going to remind us that it is officially here. December 21st, isn't it? Or is it the 22nd? I can never remember. Anyway, each year, when the first day of winter is announced, we're all supposed to groan in unison—and why not, with the prospect of three months 'till spring ahead of us? So we all groan our groans at the news and then turn back to the job of getting through the rest of the holidays.

Bah!

Have I got you in a really bad mood? Have I painted the holiday picture a convincing shade of black? Well, surprise! There's good news. I just wanted to set you up for it.

The good news is called prosperity, and it's all related to the day, later this week, that we here in America call the first day of winter. Why we call it that I can't imagine, because December 21st—or 22nd—is the *middle* day of winter. It really is. You may not have noticed, but the *true* winter season arrived without any fanfare around election day last month. So, later this week, we'll be greeting Midwinter Day, which happens also to be the first day of the true year.

It's the day the sun starts its long trip north again. Even though three cold months still lie ahead, the days will begin getting longer and and sunnier by the end of this week.

Happy True Year!

Here's a true festival day, and we haven't even been told about it. Those idiots at the weather bureau, who know better, resist any kind of change. But all over the Northern Hemisphere, from Stockholm to Fairbanks, sun-people are beginning to smile. They know what's what on earth. They know how important the sun is, and they revel in its return.

Midwinter Day requires no gifts, no turkey dinners, and no office parties. All you have to do is walk outdoors and smile, and the warmth will spread. You'll start to see an always-before-unnoticed brightness in the season, and, as the weeks slip by toward early February (when True Spring begins) you'll see the whole nature-world getting brighter, too.

At this point you're supposed to be wondering what this all has to do with prosperity. Well, the connection may seem farfetched at first, but it's as solid and dependable as the earth's movements.

Obviously, most of our economic troubles have been caused by a refusal, on the part of those in power, to face the facts of life. School children know them but the old pols sure don't. Kids know all resources are limited, that many must be carefully recycled forever.

Modern history has seen a brief spurt of supergrowth in no way resembling the normal state of things on earth. Now we're paying the price: inflation, unemployment, and shortages of all kinds.

In this twentieth-century spree of ours we've managed to flush, dump, erode, lose, or incinerate enough precious materials to keep a sane society going forever. Many of those materials now lie, unrecoverable, at the bottom of the sea. Others just fell into cracks or blew away into the sky, and here we are on another Christmas binge, knowing it's all wrong, but going through the motions because we know no other way.

History has a huge gong and some flashing lights that go off each time mankind enters another great era. They went off when we stopped being nomads and invented farming. Then they gathered dust for a while—for thousands of years, in fact—until the age of science came along. Then they began to ring at closer and closer intervals as we swept, like spoiled children, through the ages of medicine, electricity, communications, electronics, the bomb, data processing, space flight, and now . . . what?

Now we've got to grow up.

We live in an engine, called life, that runs entirely on sun-power. Every bit of food we eat, air we breathe, fuel we burn—wind, rain, cockroaches, movie stars—all were made on the sun-powered machine. Still, we talk about solar energy as if it were one of many choices open to us when in fact everything around us, and we ourselves, run on nothing but sunlight. (Unplug *that* machine and see what happens!)

Now let's go back to Midwinter Day. It's the key, in a way, to whether or not we're going to pull out of our mess. As long as we say, "Oh yeah, Midwinter Day; so what?", we'll be heading straight for famine, terror, and nuclear disaster. But the minute we get down on our knees and bless that traveling old sun, the gong will gong and the lights will flash, for we will have entered the era of prosperity, when science will serve rather than thwart natural processes, and our economy will be built on the only kind of wealth that never goes bankrupt: sunlight.

Happy True Year!

A Suburbanite's Dream

By Lorrie Otto

If we ever get over our obsession with lawn grass, this teacher-naturalist must get a big share of the credit. I first heard of Lorrie Otto when reports began to filter out of Wisconsin that a wildflower woman was helping people win court battles against laws that required lawns where native plants should have grown.

Government officials warn us that eating Lake Michigan fish more than once a week may be hazardous to our health. The reason given is that chemical contamination may exceed standards established by the Food and Drug Administration. Wisconsin has a stocking program that attempts to replenish populations of indigenous fish as well as to introduce exotic species, such as Pacific and Atlantic salmon. Fishermen say that the table quality of the salmon and trout is marginal; unless you smoke the fish, they are not all that good.

There are several reasons for such an alarming condition in the sixth largest lake in the world, a freshwater area that covers 22,400 square miles. One cause is that spawning streams have been altered, and the waterways of the country are being contaminated by "nonpoint sources" of pollution. The latter is the environmentalists' vernacular for general run-off of rainwater from rural and urban land.

Last summer I began to wonder what a suburban homeowner could do to help alleviate the problem. People caused this, perhaps people could correct it. For 25 years I have lived beside a spring-fed navigable stream that long ago was named Fish Creek because of its many fish. My two children once caught crayfish and polliwogs there. Later they netted smelt and speared white suckers in the springtime, and during the summer months my son fished for pan fish with a cane pole. When he was older, he cast for trout. Belted kingfishers and green herons screamed and squawked as they, too, took living creatures from that little creek. The children have grown up and left, as children must, but the birds have gone, too, and that has been a cause of concern to me.

Last year our village officials, contrary to the advice of state planners, allowed a developer to destroy a critical environmental corridor that bordered a section of Fish Creek. For hundreds of years, that corridor—35 acres of land—had acted as a sponge and filter for the rain, replenishing the water table and the springs. Now the land is scraped raw and awaits the builders and lawn-makers. After each shower the stream is opaque with sediment. If the smelt ran in there this spring, their gills must have been scoured by the grit. Species of fish that attached eggs to the bottom of the stream have long since had their eggs smothered. If any eggs did survive to hatch, they would not have had anything to eat, because the minute plant and animal organisms that tiny fish feed on would also have perished.

However, the developer's unusual load of sediment contributes to only a portion of the degradation. Our "gold coast" community, which is nestled along the shore of Lake Michigan in the northern section of Milwaukee county, uses the creek as a storm sewer. Pipes under sodded lawns conduct drainage from roof tops and sump pumps to ditches with concrete bottoms. Such channelization not only results in a concentration and magnification of pollutants, but also it causes flash flooding in the ravines. The water quickly rises and tears out the soil from the roots of the trees along the banks. As the trees topple and dam the creek, the next storm sends water higher around the dams, eroding the sides of the ravine and taking ever more soil and plant material out into the lake.

The bottom of the creek is becoming choked with algae as the storm overflow from sanitary sewers adds its nutrients and bacteria. Lawns are raked in the fall, and the leaves are dumped in the lowest spot in the village. Here the rains and melting snows leach the phosphorus out of the old leaves and transport this rich fertilizer into the stream. In the spring, when lawn chemicals are applied, some of them flow off into the surrounding ditches. On the land, these substances are called "fertilizer," but in the water they are called "pollutants" because they stimulate the growth of algae, which die and use the oxygen in the water as they decompose.

Added to all the sediment, sewerage, phosphorus, nitrogen, pesticides, and herbicides, are the pollutants that come from driveways and streets and roof tops. These consist of oily residue from automobiles; asbestos from brakes, clutches, and tires; dust and air-pollutant particulates; and toxic heavy metals, especially lead from gasoline. All of these flow into our streams all over the country.

Can people change this? Wouldn't non-point-source pollution be greatly reduced in suburbia if we would replace lawns with native vegetation to create a natural American landscape once again? Part of the responsibility of owning land might well be that the owner must retain and use the first inch of rainwater that falls upon his property. It is that first-flush effect that causes a sudden high concentration of pollutants to be washed into lakes and rivers early in a storm. Sump pumps and rain spouts could drain into terraced yards.. Swales, mounds, and planted settling ponds could be covered with wildflowers, numerous tiny trees, and shrubs. Even terraced roof tops have been treated this way. The deep roots of prairie plants mechanically hold soil in place as well as fill it with millions of root fibers and then store and recycle the rains. Those roots, with their accompanying insect or animal activity, create air passages where water can trickle and dissolve the run-off pollutants, which can often be used as nutrients by the plants. In shady areas, undisturbed leaf-fall and humus act as filters and sponges for contaminated rain.

Vegetation that has evolved in our climate and soils needs no additional watering or fertilizer. A diversity of plant life is not subject to pest insects. On the contrary, some beneficial insects—for example, native pollinators—might be encouraged to return to our yards. When we discontinue the use of chemical fertilizers, herbicides, and pesticides, we not only save the fish, the birds, and the butterflies and other wildlife, but also we save the energy that is required to produce those chemical fertililizers. Ammonia, a major ingredient in commercial nitrogen fertilizers, is made of nitrogen from the atmosphere and hydrogen from petroleum. Atmospheric nitrogen exists in a tightly bound molecule of two nitrogen atoms, which must be broken apart before nitrogen can be used by plants and higher forms of life to make protein. Breaking this bond requires high temperatures—usually gas. And hydrogen taken from petroleum is obviously energy-intensive.

Other pollutants that are either dissolved in rainwater or bound onto airborne particulate matter and washed into ditches can be controlled by plantings. Drainage areas along roadsides can be made much wider and deeper, even several feet *below* the drain-off point. These long swales or holding basins can be planted with vegetation, which slows the force of the water, settling out the soil while soaking up some of the moisture. We will be creating a poor man's marsh. Culverts will be less likely to clog, and the resulting wildlife corridor can be managed to display seasonal plant colors and textures. Buttercups and blue flags, purple milkweeds and rosy Joe-pyes, yellow swamps candles and helenium, reeds, cat-tails, asters, angelica, lobelia, gentians, penstemon, and so many more flowers could be planted there. We are limited only by our own ignorance! We need ordinances preserving natural vegetation in drainage easements, along swales and creeks and artificial ponds. By retaining water *on the land*, we replenish groundwater, prevent siltation and pollution of aquatic animal habitat, and also take the storm burden away from the sewage system. Many of these systems work well when it's not raining.

The little stream that goes past my home has no legal guardian, and I have no legal power to protect it. But I do have a plan, or at least a dream, that would help protect many streams. If suburbia were landscaped with meadows, prairies, thickets, or forests, or with combinations of these, then the water would sparkle, fish would be good to eat again, birds would sing, and human spirits would soar.

Wilt

By Malcolm Wells

Weeds seem to hold their shapes the way big balloons do: by a pressure from inside; water pressure in the weeds, air pressure in the balloons. A balloon shrivels and collapses when you let the air out of it. Mow a weed, or pull it by its roots, and within a few minutes it looks as if a steamroller had run over it. It dries out in no time.

The same thing happens with many kinds of flowers. If you don't put them in water pretty quickly, they'll droop over the edge like old socks. Plants seem to be the very essence of weakness and vulnerability in this tough, fast-paced world of ours. The leaves on fallen branches seem to go limp even before the storm has passed. And the words "to sap" bring to mind the idea of weakening by dehydration.

No wonder we've lost our respect for the green world. It seems to need frequent rains, or lawn sprinklings, or vases of water, to keep it from wilting, while we, on the other hand, can keep our bodies firm, and our miraculous computer-like brains alert, for weeks, if necessary, without ever going near a pool, a bathtub, or a thunderstorm.

So be it; only the tough survive. It's pretty nice to be a human being when you consider the problems of vegetation.

Just don't get caught out in the desert or far at sea without a supply of drinking water on hand, and you'll be able to maintain this illusion all your life.

Ah, the poor, drooping, faded, shriveled, spoiled, wasted, wilting, withered flowers!

But I wonder what flowers would think of things like drinking fountains and Coke machines, water coolers and canteens, soda pop, fruit juice, cocktails, highballs, beer, wine, coffee, tea, and milk. I wonder how all our drinking and urinating, our sipping and our sweating, and our noisy humid breathing would seem to a flower or a tree that quietly pulls clean moisture from the earth, and releases it—just as clean—to the open air. I wonder how quickly I'd wilt if I got no water. How long would I last on a sunny day in the Arizona desert? Is it six hours—or sixteen—before you collapse out there, and start turning into a strip of bacon under that merciless sun?

When I think of all the sipping and slurping we must do in order to stay alive it reminds me how precious rain and rivers are. It reminds me of those idiots we elected, playing with our water resources. And it reminds me of how closely I'm related to every other living creature.

I know there are a few things, like ticks—or seeds—that can go for years at a stretch without water, but the animals and plants with which I do most of my business need to drink just as often as I do. Cattle and spinach and crab grass—all are tied to the ancient ocean by the same wet strings as mine.

Does all this make you a little thirsty? Would a tall cool one hit the spot, just about now? Well, if it would, don't put it off too long or all those bright little lights in your on-board computer will start to wink out, one by one, never to reappear.

Red Buttons

By Malcolm Wells

What do you think would happen if we got a big chain—I mean a BIG chain, one with links the size of houses—and dragged it across every city, suburb, and farm in the world? Can you imagine it? Can you hear the faraway crumbling roar, like freight trains in a tornado, then the shuddering earth-tremors, as the links drew near? Fires, explosions, crashing walls, splintering trees, screams. . . .

I don't know where we'd get a helicopter big enough to lift the ends of the chain, but let's not worry about that right now; I'm just trying to picture the consequences of the plowing itself. Think what a sight it would be to watch Manhattan go! Tall buildings crumbling into unbelievable firestorms and noise, ruptured mains and chemical tanks pouring out their vile secrets: nerve gas, insecticides, virulent diseases. Whenever a nuclear power plant got caught in the chain, the massive shell would resist for a moment before crumbling into boiling clouds of radioactive steam.

Depending on the slope, weather, and soil, the death-zone caused by each of the poisons might range from a few hundred square yards to hundreds, even thousands, of square miles for the nukes. And, after it was all over, what? Smoking ruins, millions dead from side effects alone, from the poisons and exposure; billions homeless, killing for water, for food, or for shelter. Plagues. Catastrophic flooding of all the silt-choked rivers. Dust storms. And a barren, rutted land. Only a lucky few would survive.

Then, on a warm spring day, a green sprout would appear. And another. And then all over the land wildflowers would begin to bloom. Even the mountain of rubble that was New York City would turn green as the seeds sleeping in its skin brought about their miracle. With no one to stop them, the great American forests and prairies would spring up around the vast poison-scars. In ten years the whole world would look as if a master landscaper had been at work. Profusions of wild plants would adorn every unpoisoned inch. Birds and animals would move in from the unplowed wilds to live on the rich, new bounty. In 50 years little sign of our civilization would remain.

And the survivors? Well, I imagine they'd be living pretty simply for a long, long time. Many would die inexplicably from the invisible poisons, but, little by little, the locations would be recognized and avoided.

Whether or not any lesson would have been learned is debatable. Chances are that in a few generations the survivors' descendants would be in all the messes we're in today. But they would be alive. Even the most miserable for us treasures life. Life would have survived. That's the point.

Now let's think about the little red buttons in the White House and in the Kremlin. Touch one of those scarlet disks and you unleash a final night of terror that would make my chain-drag horrors look like a romp. The thought of nuclear war is so awful we can't, or won't, even think about it. Instead, we put ourselves in the hands of our leaders and hope for the best.

Maybe I'm wrong, but I think we need the peace movement now more than we ever did in the Sixties. World politics never looked more unstable to me.

Nature, Naked Nature

By Malcolm Wells

Sounds like a name of a topless dancer, doesn't it? Well, it's not. The title is part of a line from Thoreau's works, which goes like this: ". . . nature, naked nature, inhumanly sincere, wasting no thought on man . . . ."

Henry Thoreau wrote those words before Lincoln was elected, but they're as true today as they ever were. Thoreau wasn't condemning nature; he was simply looking at it without all the trashy sentimentality that so often blinds sunset-watchers like me.

Nature doesn't live under Disneyland rules. The hurricane kills the good along with the bad, and the horror of a drowning goes unnoticed by the waves. I click my teeth together, never realizing that in doing so I've just killed a hundred bacteria. We can poison seas and pave continents; we can erode all the soil and bomb all the cities; but in the end we'll only kill ourselves, and nature will survive.

Believe it or not, this happens to be an essay on landscaping. The best way to design a landscape is to understand nature, simply to stand back and watch her for a while: see what the old girl is doing when no one's zapping her with trail bikes or DDT.

If I want to landscape a 50 x 50 foot plot, how many trees should I plant? (I'm not talking about starting a nursery or an orchard. I'm simply talking about a natural landscape that will be most appropriate wherever I happen to live.) Would four trees be enough? Would ten? Twenty?

Would you believe a *thousand?*

My formal, dot-dot-dot landscape-designs look hopelessly inhibited in comparison to nature's wild exuberance. Have you noticed the way young trees take over an abandoned field? They grow so close together you can hardly roll a cantaloupe between them. "Sure," you say, "but look at their mortality rates. Only the strongest ones survive."

That's the point; nature's been doing it that way for a thousand million years.

We can't *create* landscapes, anyway. All we can do is help a little. A landscape is a living community made up of thousands of different plants, animals, insects, enzymes, and nutrients. If you tried to buy them, they'd cost you a fortune, but that isn't necessary.

You can plant trees and shrubs. And you can add some rotted manure or mulch, and give everything a good sprinkle. But then go away. Each time you come back you'll see a new miracle: thousands of green tips emerging in the spring, birds appearing, butterflies, wildlife, wildflowers, changing colors . . . and the whole thing getting better and stronger and more beautiful all the time.

But there's a catch: in many towns wildflowers are prohibited by law. They really are. If you don't mow the grass you'll be fined for making weeds, and it could cost you a lot of money. Which brings us to the subject of higher laws, and makes me an advocate of lawbreaking, I guess, but only in the way Henry Thoreau advocated it: willing to pay the penalty.

At the Nuremberg Trials a world policy on higher laws was formulated. It said, in effect, that each of us has a duty to refuse orders he cannot in good conscience obey. The policy, of course, referred only to human affairs, but the principle must apply to any conflict between authorities.

It's getting pretty late in the century for us to be still tiptoeing around, intimidated by some ridiculous 1938 anti-weed ordinance passed by a politician more concerned with neatness and poison ivy than with the basic facts of life on our planet. There isn't time, anymore, to live by such subjective standards. If we don't soon back off and let nature thrive again, she's going to show us just how naked and sincere she can be.

Antipode

By Malcolm Wells

Everybody has one. And everybody's antipode is a little different from everyone else's, unless, that is, someone's standing on your shoulders or lying on top of you.

The place on the other side of the world that's exactly opposite from the place where you are is called your antipode, and it isn't in China, either, not unless you live in southern Argentina. The Philadelphia antipode is in the Indian Ocean, about 1,000 miles from the city of Perth in western Australia, so don't try to dig your way to China or you may find a shark instead of a Maoist.

Now, before these boring facts put you to sleep, I want you to look down.

Straight down through the earth beneath your feet; at this very moment, far, far away, fish are swimming upside-down in a wild, stormy ocean you may never see. Can you imagine them being there? Look down, and really think about it.

From their point of view, here, eight thousand miles below their tapered bellies, two million upside-down Philadelphians are living in filth no self-respecting fish could tolerate. Which is still a boring fact, I guess, until the reality of this thing we call Earth suddenly hits, and we see what a frighteningly lonely miracle this blue-green ball of rock really is.

Traveling forever through the emptiness of space, we sail on a ship that's already stocked with all the provisions it may ever have.

We're the rich passengers. We eat three meals a day and watch television. The other passengers get barely enough to eat, and many don't even get that. Everything we do threatens them, we're so rich and powerful. Our car, our heater, our lights, our friendly neighborhood supermarket—each uses provisions that other passengers badly need. They'd use our leftovers if we'd let them, but we send most of our garbage straight to the bottom of the sea—or into the sky—out of reach forever.

Does the President know all this? Probably. Do you think he's ever considered living a life of voluntary poverty as an example to the rest of us? I wouldn't count on it. But that's not what disturbs me most. The first question is, what have *I* done? What have *I* saved for those other passengers—or even for those antipodal fish whose fate I hold in upside-down hands so far below their gills?

To Tell the Truth

By Malcolm Wells

For several years, groups of citizens in Cherry Hill, New Jersey, complained about the tons of partially treated sewage being dumped into the river each day. Finally, after great deliberation, the city fathers took action to correct the matter: they changed the name of the Sewer Department to the Water Pollution Control Division. It solved the problem.

The U.S. military establishment may no longer be a war department, but when you really think about it, you know it's a hugely wasteful empire, richer than most nations. Ever since it was renamed The Department of Defense, however, we've all paid our monstrous taxes and gone back to sleep.

The easiest way to get critics off your back is to change the name of what you do.

I've just returned from a trip on which I visited a widely publicized "environmental" college. I'd looked forward to my visit with a mixture of dread and hope; dread of finding a project far better than anything I'd ever designed; hope of seeing, for the first time in my life, a building that didn't destroy the ground upon which it stood.

But the closer I got the worse it got. First, I saw the huge parking lots—sheets of dead asphalt where a forest had lived. Next, miles of sidewalks and plazas and paved courts and terraces. All I could think of were the flash floods all that paving created every time it rained.

Then I saw the buildings: long walls of glass to roast the kids in summer and freeze them in winter. Nowhere did I see a storm sash. Nowhere did I see a window you could open when the air conditioners broke down. Sealed tightly, the whole place was forced to depend on machines and fuel every minute of the year.

Aluminum—the energy-spending material—glittered on many surfaces. Walls had hardly any insulation. Water gushed from untended lavatory spigots. Lights blazed all day. It was winter, but great clouds of heated air were being vented away into the winds. And the food—the food in the college restaurant—consisted of soggily overcooked greasy material of unknown origin, served on throw-away plastic plates! This was the environmental college; the school of tomorrow.

I know it's hypocritical of me to condemn all this. I'm very much aware of the equally wasteful buildings I've designed, some of more recent date than I'd care to admit, but the "environmental" label on this college seemed to make it fair game for criticism. All too often we visit such places and come away remembering only the labels. We fool ourselves because it's easier—and less complicated—to live that way.

But most of us architects know better now. We know, for instance, that the production of aluminum requires massive doses of energy. We recognize the importance of really good insulation. We know about sun angles and body temperatures and porous paving. We know we should make waste-separation easy for building occupants. We know it's wrong, and monotonous, to light whole rooms uniformly when low-wattage task-lighting can illuminate only the surfaces being used. We know, too, that in the earth's eternal scheme of things, green, living plants were meant to cover as much land surface as possible. And we know that planted areas should be left wild and free, not regimented into dot-dot-dot landscapes. Still, it's so much easier to do the old standard hack-job, lable it environmental, collect our fees, and move on to the next project.

Sometimes when I think about all this I get depressed. But, each year, wildflowers continue to poke up through the asphalt, and seedlings sprout by the curbstones. Nature's optimism is so infectious I start to get infected, too. With all this beautiful green sincerity stinging us each season, who's to say we environmentalists won't soon turn sincere as well?

Yecch!

By Malcolm Wells

For a longer time than I care to remember, I have been what you might call a sensitively attuned, earth-conscious human being. An ecofreak, in other words.

You know the type: vegetarian, bicycler, beard; the whole works; a walking expression of self-righteousness; the kind that goes around turning off lights and making a great show of saving string. Lately, I've even caught myself talking to plants. I guess the next step is to wear rags—or practice meditation.

But I guess where I am right now: thirty-nine thousand feet above the earth, riding in a monstrous, fuel-gulping jet. On my tray-table there's a dry martini. And I'm smoking a cigarette.

I'm having a relapse.

My nerves are all frazzled because for the past three days I've been held captive by a group of people just like me: earth-poets, old hippies, organic gardeners. For three days I've been sitting with them, somewhere out in the wilds of America; sitting on the floor, of course, wearing my sturdy walking boots, sneering at pesticides, and deploring plastics. We've been eating so much wheat germ and being so nice to each other I can't stand us any more. But I've been too far from any transportation system to get away until now.

That's why I broke into a run the minute they called my flight. You can't imagine the relief I felt when the passenger door finally clumped shut and I could become a plain old ordinary citizen again. But don't get me wrong, my rebellion is only skin deep. The minute I get home I'll be turning off lights and lowering thermostats the same as ever. I'll still be sneering at pesticides and living on vegetables.

It's just that I've had it up to here with poets in floppy hats. And alternate life-styles. And health-food stores. I mean, why do we need *health*-food stores? Why can't the stuff we buy in the supermarkets be good food instead of colored chemicals? Why must we become weirdos if we want to change this whole mess? Who says we can't have forests and gardens in the centers of our cities? Why must suburbs have such mind-crushing monotony? Will the good life forever be available only to millionaires and to repulsive old runaway freaks in floppy hats?

I don't know. But the more I think about it the more convinced I am that we're going to get nowhere until we do it all together, as plain, ordinary, lovable, nine-to-five citizens. With a dream.

The P.V.H. Indictments

By Malcolm Wells

It's obvious that a society with as many doctors, nurses, hospitals, clinics, and drugstores as ours is not in what you'd call the very best of health. Such numbers tell a frightening story. Whether they come from hospital soap-operas or medical journals, they never fail to remind us that all is not well in the health department.

And when we consider the subject of *city-health*, the look of a slum can tell us, far better than any sociologist can, how much work we still have left to do. Stinking water is all we need to judge a river's health; burning eyes indict each smokestack in the land.

But some of the most damning evidence of environmental decay slips by, completely unnoticed, because we've been misled by it for so long. Take parks, for example; you know: bridle paths, woodlands, and picnic tables. "What's wrong with them?", you say. Well, isn't it a pretty sick world that has to protect green areas from itself? I mean the whole damned continent was a park before we got here.

Parks

Parks are like zoos: jails for nature, places where you can go to see beautiful things we'd kill if they ever took down all the fences. What kind of monsters are we, anyway, to be so destructive that we have to protect parts of our own world from ourselves?

Vacations

Or vacations; what are they but expensive bits of pleasure set aside in two- or three-week packages to make the other fifty weeks more bearable? Can you imagine how we'd look to a creature from another planet, having to run away from our lives, once a year, in order to stand them? Where in the rest of nature—when in the rest of history—has the vacation idea appeared? There's nothing wrong with a change of scene, but if that's all we're after, why don't we all go visit each other's factories and offices when vacation-time rolls around? No, it's all the artificiality of modern life that crushes us. We crave the real, living world of nature.

Health Foods

And health-foods; what an indictment they are: what kind of junk must supermarkets sell to make the health-food industry boom the way it does? I mean why do we need health-foods, anyway? Why can't food-foods be good for us? How do food-processors get away with what they do? You'd think they were secretly in league with the doctors.

We have to keep reminding ourselves that the finest, purest vitamins, minerals, and nutrients don't come in pill bottles. They come in food—in natural, undoctored, left-alone food—before all the bleaches and tenderizers and preservatives are added.

The health-food stores get rich, the vacation industry dry-cleans our bank accounts each year, and we let park-like forests and fields turn into these deadly shopping centers. But if we ever got angry and decided to make everyday things, like food and work and cities, pleasant and healthful again, then parks, and vacations, and health-foods would become freaks of history; half-forgotten aberrations in our long, zig-zag journey.

J.F. Mamjjasond

By Malcolm Wells

Every year, J.F. Mamjjasond comes around and makes me breathe in and out 10 or 12 million times without stopping. Then I have to eat 1,100 meals before she comes 'round again. Which sounds harder than it really is. Almost everyone has to do it.

You know who Smut Wutfis is, of course, and Prendel, and Qwerty Youie-opp. But how often do you think about J.F. Mamjjasond? I see her name so often I can't help thinking about her. And besides, her visits now seem to come so often I couldn't forget her if I tried.

(What's that? You don't know who Smut Wutfis, Prendel and Qwerty Youie-opp are? Shame on you. Smut Wutfis is printed across the top of every page in my date book. SMTWTFS. Prendel is the message printed above most steering wheels, at least in cars with automatic transmissions. PRNDL. And Qwerty Youie-opp dances across the keys of every typewriter. QWERTYUIOP. That should help you identify J.F. Mamjjasond. Her name's on every calendar.)

J.F. Mamjjasond has been around a long time. Around and around and around, ever since someone decided that years should be divided into months. Before that it was just year followed by year followed by year. When mankind finally disappears, J.F. Mamjjasond will disappear too, along with Smut Wutfis, Prendel and old Qwert. They remind me of the King brothers, Nosmo and Nopar; so powerful when backed by municipal ordinances, but who, like J. F. and Qwerty, are but puffs of air without us word-users here to see them.

Can you imagine being a bird, having to take long trips without packing as much as a toothbrush, or buying flight insurance? Where are the directional signs in the woods? How do a million different species go about their business, obey their laws, and avoid collisions without signs or instructions?

You'd think we'd be lost forever if we had to live like that, but when the signs disappear inner voices are heard. We walk around, battered by suggestions and instructions—stop, go, eat, drink; do this, do that— largely unaware of those more powerful instructions printed in our genes.

The world races along its invisible path while we enhance our interdisciplinary methodologies with interfaces and parameters, gobbledegooking our way through time.

Do you suppose anyone is out there watching us, and smiling just a little?

Waste Made America What It Is

By Russell Baker

When he's not gracing the cover of Time, *Russell Baker writes a column for* The New York Times. *But what's a humorist doing here? Being deadly serious, that's what he's doing, deadly serious about a big part of our energy trouble.*

The United States is the home office of waste and always has been. The country was built on waste. We wasted land, wasted people, wasted resources, and wasted fortunes that were built on wasting land, people, and resources. Large parts of the country stretching from the South Bronx to Los Angeles shopping centers are now pure wastelands.

I respect waste. Waste made America what it is today. Some people talk about the Constitution, some about the Conestoga Wagon, some about the Colt 45 and some about the railroads, and, indeed, all played their role in building the country. But what good would they have done without waste?

It behooves Americans to cherish waste as part of our heritage. Yet, what began a few weeks ago as a decent old-fashioned tax revolt has now degraded into an attack on waste.

A poll conducted by *The New York Times* and CBS News suggests that three of every four citizens want governments to stop waste. Most of these people want to have their taxes cut, which is natural, but don't want to lose any government services they now enjoy, which is equally natural.

The question with which they then struggle is how governments can take in less money without reducing services. The answer is to end waste, which is thought to be rampant in government, and probably is.

The trouble with this solution is that it offends the American character. Any government that did not practice waste on the grand scale would be a poor representation of the American people.

Here let me now make a confession. I have just thrown out a half-bottle of carbonated water. It had gone flat because I forgot to put a stopper in it. I wasted that water and did not feel the smallest pang of guilt about doing what government does every day.

What's more, I wasted the bottle it came in, and I wouldn't be surprised if the bottle cost more than the water I wasted. The company that makes these bottles insists that I waste them. So does the union that works for the company that makes the bottles. Both management and labor believe that waste is good for business.

This is not an isolated case. The supermarket is selling tomatoes lovingly wrapped in molded plastic. You know that costs something, but what do you do with the plastic after eating the tomato? You waste it, right along with the paper bags in which the supermarket packages the carbonated water with the bottles designed to be wasted and the tomato wrapping, which isn't fit for anything but wasting.

I do not quarrel unduly even with such conspicuous waste. Cannier people than I, people who are geniuses of commercial enterprise, find waste a profitable undertaking, and I respect their judgment. I buy their cars, designed to be wasted, and their pens, built to be thrown away. What puzzles me is why they expect government to behave differently from the people who buy their merchandise.

We had some people in to dinner the other night and they got going on the terrible taxes and from there took off against the evils of government waste. In fact, they became so absorbed in deploring waste that most of them neglected dessert, thus wasting the better part of a strawberry pie.

Most of them, like me, drive automobiles whenever the impulse strikes, thus contributing to the national wasting of gasoline, which has led to the decline of the dollar. None of us worked up any heat against ourselves for indulging in this traditional American waste. In fact, I didn't even hear it mentioned.

Anybody who did mention it would've been considered a bore. It's boring of me to mention it now. We all looked energy-wasting squarely in the eye a long time ago and decided to keep it as part of the American tradition. Right now, I'll bet, there are thousands of people writing letters to editors and congressmen about the viciousness of waste while running air-conditioners that waste power at a prodigious rate.

Wasting power on air-conditioning is an accepted American tradition, even if Abraham Lincoln did do his letter-writing with only a cardboard fan to cool his brow. The chances of persuading governments to stop waste are probably not a bit better than the chances of forcing ourselves to kick the habit. If the tax revolt deteriorates into an attempt to make governments behave more sensibly than we do, it is surely doomed. Success depends upon forcing government to do less for us.

If we truly want lower taxes, we shall have to learn to back politicians into the corner and snarl, "What's the idea of trying to do something for me lately?"

The Birds
of Bellazon

By Malcolm Wells

Jersey is what most of us call that paradise of asphalt and neon lying between Philadelphia and New York. But there's another Jersey, an island off the coast of France. That other Jersey is about twice the size of Manhattan, which makes it pretty small for an ocean island, but with its population of only 100,000 people, most of whom are crowded into its quaint coastal towns, Jersey has room for hundreds of little stone-walled farms and miles of wild seacoast. It's quite a place.

Being an island and being dependent on tourism and agriculture, Jersey can afford neither the ugliness nor the land destruction that municipal dumps involve. So, when the island trash problem began to get out of hand 10 or 15 years ago, the people in charge, being a different breed, apparently, from U.S. politicians, decided to find an innovative way to turn their liabilities into assets. Maybe it's just my rose-colored tourist-glasses and my Watergate cynicism, but politics seems more of an honorable profession over there.

Anyway, when they began to run out of trash-dumping space, the Jerseymen looked at all the alternatives and decided to follow nature's ancient example: return the wastes to the land. In a hidden valley, they built a processing plant that not only turns trash, garbage, and sewage into organic mulch but also runs itself—motors, lights, and all—on electricity generated by the marsh gas that bubbles out of the sewage.

Not all trash can be composted, of course; some must be recycled. Jersey's bulk cardboard and paper are sold to London papermakers, and the metal and glass are sold as scrap. The plastics are incinerated. That's the only sad part—the incineration—even though it's minor compared to the mountains of wastes that *aren't* incinerated. But it is sad. On the day I visited that valley I saw smoke before I saw anything else.

Bellazon Valley is narrow, wide enough for only a tiny stream, a strip of forest, and a road. Walking toward that valley from Jersey's only city, I was soon out among the tiny farms. I congratulated the Jersey men on the fact that one of the world's best waste-management plants was located in one of the world's nicest spots. Then I saw smoke straining the brilliant sky. For the next quarter mile I could see nothing of the plant but a giant chimney above the trees—and seagulls, hundreds of them, flying in and out of the smoke, excited by something I couldn't see. Then the plant came into view.

Well, it's no architect's dream, but then what would you expect? It looks like a cell block for the outcasts of Jersey, and that's exactly what it is. The gull-excitement was caused by all the food scraps in the garbage. The birds were stealing it right off the trucks, and they were screaming their heads off telling me about it.

I won't bore you with all the details of the Bellazon plant. Just take my word that it is impressive. Tin cans leap out of the trash and ride away on magnetic conveyors. Big trash is pulverized into little trash. And treated sewage-sludge is vacuum-dried on strange machines and turned into fragrant flakes.

Quite a sight.

Then the whole mixture, tons and tons of it every hour, is dumped into the top of a six-story building, where the heat of natural composting turns the stuff into useful land-mulch. Each day's mixture is automatically turned and dropped to the next level until, on the sixth day, it falls into the waiting trucks of happy farmers. How do I know they're happy? They must be; they wait in line and pay for the stuff. They can't wreck their farms with government-subsidized chemical fertilizers the way we can. Some farmers buy straight sewage sludge; others buy the full mixture.

Now you know all about the waste-management plant in Bellazon Valley, but I suspect the full relevance of the story to your own town may still be obscure, so let me explain: your town is in deep trouble. It suffers along with mine from over-richness, a disease our troubled economy may slowly cure. But for years to come we'll continue to throw away incredible amounts of waste materials while others starve.

Chances are the people in your town still rob the land by putting lawn cuttings and autumn leaves in plastic bags for collection along with other "wastes." Chances are your own weekly trash could fill the back of a Volkswagen. A lot of the blame must go to the people who produce double wrappings and junk mail, but the fact remains we all waste wastes.

The Isle-of-Jersey process isn't the final answer, but just think of the promise it holds! Everywhere that you see desolation, unemployment, filthy rivers, and city dumps you could be seeing lush, green land, gardens, gardeners, and crystal waters again, simply by turning wastes into riches as the Jerseymen do.

Think what a hero your mayor would be if he closed the city dump, turned wastes into compost, and sold it to reduce your taxes! Think about the beautiful waste-processing center that would house the process: a manmade park—trash and sewage going in one end, wealth and jobs coming out the other. It will happen, you know, when we care enough. All you have to do is put the squeeze on city hall.

Then you can tell me about the happy birds where you live.

Throw Away the Keys

By Irwin Spetgang

When I write solar books, I write them with Irwin Spetgang. His are the technical brains of our little solar company. I do the dummy stuff and illustrations. Here's Irwin on a favorite subject.

Occasionally my anger flares, sometimes so that I act impulsively. Once in a rare while it becomes an all-encompassing rage, and my reactions surprise even me.

One such incident occurred about fifteen years ago as I was driving north on a major Philadelphia artery. My wife and two young daughters were with me in our VW Beetle when the driver of a large sedan nearly stopped our world. He was zipping past us on the right, when, to avoid traffic in his path, he cut abruptly into our lane. Fortunately, there was a break in the southbound traffic at that instant, and I was able to avoid the collision by veering to the left.

As I brought my VW Bug back, a wash of relief swept through my body, but my anger began to grow. Our assailant, now several car-lengths ahead of me, had nearly mutilated most of what I hold dear in this life. The thoughts of how vulnerable we were in our small car, and that his casual carelessness could have made our world a nightmare, caused my anger to grow into outrage!

Then it happened. A sweet quirk of fate. A red light halted traffic ahead of my tormentor and he was forced to stop—directly in front of me. Despite my family's vigorous protests, I jumped from our car and ran to his. While venting my anger with choice vocabulary, I opened the door to his car, reached in, and snatched the keys from his ignition. (I didn't even consider that he might be bigger and stronger than I, a frightening thought in retrospect.) Returning to my car I turned and threw his keys as far as I could, while nearby drivers who had witnessed the whole tableau offered approving comments and applause for what they obviously thought to be Just Retribution.

A fantasy? No. It really happened.

I believe the person in the other car is trying to do you in! Not deliberately, but with the same casual thoughtlessness as the driver in my Philadelphia encounter. You know him. He's sitting in the parked car with its engine running. Often you see him double-park as he leaves his vehicle (always with the engine running) for a quick stop at the post office or drugstore.

"So what?" I hear you say. "Isn't it just as fuel wasteful to stop an engine, then start it again in a few minutes later?" The question annoys me because I'm not sure of the answer. With this doubt nibbling at my righteous indignation, I decide to do some homework. I call the offices of General Motors, Chrysler Corporation, and American Motors.

The first thing I learn is that it's not as simple as asking "How long can a car idle before it begins wasting fuel?" Bill Winters of GM throws questions at me: "What model cars are you referring to?" "Are they well tuned?" "What size engines?" "How old are the vehicles?" His machine-gun response to my question almost has me on the run. I hurl back at him: "Midsized passenger car, six-cylinder gasoline engine, in good repair, two years old." (My gut feeling is that I've defined some average car.) With this return salvo, he lets down his guard and says he's never had a request like mine before, but he will get back to me with the information.

I don't hear from Mr. Winters for about a week, but when I do his answer is gratifying. He tells me that Dr. Craig Marks of the GM Engineering Staff says that you're wasting fuel if you let your car idle for more than about 30 seconds!

Lloyd Northard, in the American Motors PR Department remembers a booklet that addresses the question of fuel use in idling autos. "When parked or stopped, if only briefly, the ignition should be turned off. Idling requires a rich fuel mixture—about one-half gallon of gasoline per hour is consumed. One minute of idling will consume more gasoline than is required for restarting."

Now we're getting somewhere. I begin to get that warm glow that comes when my intuition has the last laugh. Would the Chrysler Corporation let me down? Tom Jakobowski, their spokesman, tells me he doesn't know the answer, but he feels that it is common sense to turn off the ignition when waiting in a parked or stopped car. He checks for more accurate information and calls me back. A small car, six cylinders or less, should be turned off if you must idle for more than about 15 seconds. If you're in a larger eight-cylinder model, don't let it idle for more than about 20 to 30 seconds. Before that, it takes more gas to restart the car.

But I don't need the final details, because his first response is the strongest argument.

"Common sense," he says.

Do those abusers who leave their engines running lack it? Or is it that they just couldn't care less?

In examining my anger at them, I wonder if I see villains behind every tree, or if it is a real problem that inflames my passions. So I take inventory—at the curb in my local shopping center, at the malls, in the downtown business district, and at the local convenience "mini-market." In fact I look everywhere that the VIP might leave his car idling. The answer is consistently the same. They *are* there. Fouling our air and wasting our fuel.

The scope of their abuse begins to sink in. Rich fuel mixture, about one-half gallon of gasoline an hour. One car here, one there, perhaps ten or twenty at any given moment in any given town. And hundreds and thousands of towns and cities.

It *does* make a difference! A little awareness could help buy the extra time needed for our energy-gulping economy to shift gears, so its needs can be satisfied in other ways. Is there anyone in today's world who is not aware of the fuel shortage haunting mankind's future? Yet, these drivers seem not to care when they waste the rapidly diminishing asset that will be more valuable to future generations than gold is to ours.

Perhaps even worse is their polluting of our air with that insidious accumulation of poisons coating the lungs of all living creatures, and dimming our views of a crystal sky.

Why hasn't my anger grown into a rage? Is it because the attack of these assailants is so slow-acting that the danger holds no immediate threat?

Or is it that I am a coward, that I fear walking up to the stranger and telling him to switch off the ignition of his idling car?

When will this smoldering anger in my chest bubble over? When will it become the outrage that will let me reach into his car and throw away the keys?

Here's a card Irwin Spetgang has had printed:

PLEASE TURN OFF YOUR IGNITION

Though you may not realize it, you are:
- Wasting fuel, and
- Polluting the air needlessly.

Your cooperation is appreciated.

(After 15-30 seconds of idling, you've used more fuel than it takes to start your car again.)

Why not obtain a supply of these cards and hand out a few?

Fear of Lying

By Malcolm Wells

*To Captain W.L. Guthrie of EAL, who taught me to see smoke
through all the haze from Washington.*

*Dashing through the smoke, in a brand-new jumbo jet,
Over hills of oak. Haven't hit one yet.
O'er the fields we go, laughing all the way
As we try to hide our prayer,
"Please don't crash today!"*

*Denver smokestacks belch. Boston tailpipes blurt.
Albuquerque power plants do the coal-smoke squirt.
Everywhere you look: murky shapes you see;
Vistas lost in brownish gook.
Will a plane hit me?*

"Good afternoon, ladies and gentlemen, this is the captain speaking. At
the present moment we're passing directly over Bismarck, North Dakota, and those of you sitting on the right side of the cabin should be able to
see the Missouri River in a moment or two. We're flying at an altitude of
42,000 feet, the outside temperature is a brisk 37 degrees below zero,
and the weather is clear, with a visibility of 25 miles."

No one notices the mild-looking man who starts to scribble a message as the announcement ends. When he rings for the stewardess only a few heads turn. But when he sends her forward with a note for the captain, a sudden ripple of terror spreads along the aisle.

The stewardess returns, hands the man another piece of paper, which he reads. He shakes his head, and slips the note into his pocket.

"What in the hell is going on?" people wonder, every time the scene is re-enacted.

Well, I'll tell you: I send a note to the captain every time I fly, hoping that someday one of those gifted men will be moved by my eloquence to amend his anouncement.

Here's what my note usually says: "Sir, Why do you call the sky 'clear' when the horizon is totally obscured by smoke? Isn't clean air transparent? Shouldn't we be able to see every detail of the landscape for hundreds of miles from this altitude? Don't you ever want to cry for help against the shiny metal airplane that's going to come roaring out of this glurk one of these days and sprinkle little bits of us all over North Dakota? Don't you know that most of us, back here, think we're looking at a natural atmospheric condition when we see all that smoke outside?"

And here's what usually gets handed back to me: "Dear Mr. Wells, Thanks very much for your interesting comments. The safety measures aboard these aircraft make horizon-visibility unnecessary. You can be sure that we in the industry are doing everything possible to make your trip safe and pleasant."

What fun it is to ride and sing a slaying song tonight.

Waterbags

By Malcolm Wells

TWA has a regularly-scheduled flight to California that can carry 200 giant plastic bags full of warm water. Each bag holds about 15 gallons, plus a few pounds of structural materials and chemicals. The bags get flown across the country to be exposed for a few days, or a few weeks, to similar bags of water and chemicals. Then all the bags get hauled back to Philadelphia again!

They have to be kept from freezing, of course. That's obvious. What isn't so obvious is the extra care those bags require. They must be kept at a carefully controlled temperature, they must be fastened into specially-shaped airfoam cushions, and they must be given in-flight doses of additional chemicals. During each flight many of the bags threaten to overflow and must be accommodated by all sorts of special piping and drains.

"Well, what's wrong with that?" you may ask. "It provides jobs for airline personnel and airframe builders, for ground transportation people, and for a lot of related industries, not the least of which is the petroleum industry."

And there you have the catch. The energy crisis may be pushed back in your mind, but there's no question that the world's supply of oil can last for only a few more decades. We've got to start asking ourselves if we shouldn't save some of that precious black lubricant for the trillions of descendants who still lie sleeping in our genes.

Nature is not making oil anymore.

Not any.

That's why the hauling of those waterbags all over the country at jet speeds can hardly be justified except in the most critical of emergencies. The waterbags I've been talking about, as I'm sure you've guessed by now, are people.

I love to fly, and I'm going to resist being grounded as much as the next guy. But America has been in business for over 200 years. It's time we made mature decisions. We know we must stay closer to home, we must give up our big cars, and we must make wider use of local transit systems. The question is, which kinds? Buses? Trains? Moving sidewalks? None are ready, on a country-wide basis, to serve the cities and suburbs.

There's only one all-weather system that's perfected and ready to go. It's free, it's healthful, it's available to most of us, and it improves with use. I'm talking about walking.

Now wait; don't go away. This isn't just a whim. All the facts imply that we *must* use that system. Look at it this way: over half of your entire body is there strictly for the purposes of walking and running. That means that a large part of all the food you buy is used to fuel a transportation system you seldom use. Add *that* to the cost of owning your car! You're keeping your body's own car locked in its garage, so to speak, at great cost in terms of food bills—and heart attacks—while ever more inefficient travel devices replace your legs.

Is that what you really want? If we're going to replace the leg shouldn't its replacement at least be efficient?

That's where I stand.

The Flesh Is Always Willing, So Harness It

By George Sheehan

*Every runner knows of George Sheehan. Most have read his books. He's
the physician-philosopher for the people you see bouncing along the sides
of streets. "At 60," says this keeper and reader of journals (Thoreau,
Emerson, Kierkegaard, Ortega), "I have broken out into a new life. It is
filled with urgency and the awareness of a potential yet to be tapped.
Like Frost, my fear now is that my offering will not be acceptable."*

I was on a radio show recently, discussing exercise with a woman who
did not exercise. "The spirit is willing," she told me, "but the flesh is
weak."

I had, of course, heard that excuse many times before. But for the
first time, it occurred that the opposite was true.

The flesh is willing. It is the spirit that is weak. Our bodies are
capable of the most astounding feats. But the horizons of our spirits do
not reach beyond the TV, the stereo, and the car in the garage.

The flesh is not only willing, it is wanting and waiting to be in
action. The flesh is filled with everything the spirit lacks: grit and pluck,
and nerve and determination. We come from a breed that crossed
continents on foot, and trekked from pole to pole. And even now we see
housewives running marathons, stockbrokers in Outward Bound, re-
tired executives climbing Everest.

109

We are of a flesh that asks for more and more challenges, that seeks one frontier after another. What is missing is not physical energy. Physical energy is there for the using. The fuel is there. It is waiting to be ignited. We need something to light the fire, something to get us into action.

We can see this from the moment we wake up in the morning. We lie abed awaiting the third, and last, and now frantic, call. The alarm clock, the radio, and the family have taken turns trying to get us up. Still, we lie immobile until the last possible minute.

Survey this scene and tell me that the spirit is willing but the flesh is weak. How many calories does it take to get out of bed? Whose bodies are so exhausted that they can't get their feet on the floor?

I can plead that I'm in a semi-coma, not yet ready for coordinated action, but the same inertia happens again and again throughout the day. The body is ready and willing and able. The spirit is becalmed. Where there is no emotion, there is no motion, either.

What is missing is the spiritual energy, what the Greeks called "enthusiasm." There are, of course, many other desirable qualities missing as well, but enthusiasm is the key.

It is from lack of enthusiasm that the failures of the spirit multiply during the day. We must, as the word implies, be filled with, and possessed and inspired by, a divine power or spirit.

When we are enthusiastic, we take on those qualities that go with it. We develop a determination to equal the endurance of our muscles, a fortitude to match the courage of our hearts, and a passion to join with the animal strengths of our bodies.

To succeed at anything, you need passion. You have to be a bit of a fanatic. If you would move anyone to action, you must first be moved yourself. To instigate, said Emerson, you must first be instigated. I am aware of this every time I lecture.

For an hour before the talk, I can be seen walking alone, muttering to myself, gradually building myself to a fever pitch. So I find it completely natural to end a talk standing on a table with nothing on but my Levis, and with them rolled to the knees.

But the spirit has more to offer than just this excitement. It gives us motivation and incentive when the excitement is missing. The spirit is what gets us through when everything else fails. In his paper on "Factors in Human Endurance," Oxford professor Ralph Johnson points this out. A man's ability to survive, he states, depends on the qualities of his personality. This thought is particularly striking in the accounts of explorers and mountain climbers, people stretched to their limits and beyond.

The explorer, Captain Scott, writing of one of his men, commented, "Browers came through the best. Never was there such a sturdy, active, undefeated man." Of Scott himself one of his companions wrote, "Scott was the strongest combination of a strong man in a strong body that I have ever known. And this because he was weak. He conquered his weaker self and became the strong leader we went to follow and came to love."

So behind the enthusiasm, behind the inspiration, behind the passion, there must be the will. What finally and irrevocably separates us from the rest of the world is our will.

We can choose. We can decide. We can will to do it our way. And when we do, nothing can prevail against us.

Otherwise, we are merely wishing. We are in the world of the lukewarm. "The spirit is willing but the flesh is weak," is the cry of the lukewarm, the lukewarm in anything.

The spirit is willing, I might say, but my brain is weak. The spirit is willing, I may tell you, but I lack the social assets. The spirit is willing, you see, and it's not my fault. No wonder we use the word wishy-washy.

We must want something, and want it badly. Want it with a zeal and passion and enthusiasm of a Don Quixote or a foreign missionary. Then we will suddenly be in motion, focussing on that goal.

Once moved, the spirit and the flesh are like a matched team of horses, each asking more of the other. And in the end, they fuse, so that for a brief and wonderful moment we are filled with the divine—when the flesh and the will and the spirit become one.

I Gotta Fire Inside A' Me!

By Warren Stetzel

Warren Stetzel was active in the civil-rights movement as far back as 1950, before I and most others had even heard of it. Typical. I caught up with him in 1970, not long after he'd helped organize a group—Raven Rocks—that managed to snatch a beautiful Ohio tract out of the jaws of the strip miners and is now building a highly experimental multifamily house and office on the site. Teacher, author, and for five years a close associate of historian-philosopher Gerald Heard, Warren makes things happen because he cares—and dares—enough to plunge in and try them.

It was a frigid January morning in Ohio. And there he was, little Jimmy, out on the walk with no coat, no boots, no cap, no gloves.

"You're gonna get cold, Jimmy!"

"Nah!" he insisted. He stopped and looked up at me. He was positively exhuberant. "I gotta FIRE inside a' me!" He puffed a cloud of steam into the frosty air, and took off.

Jimmy had looked me squarely in the eye. What could I have done, but take him at his word? He looked warm, as a matter of fact, and it wasn't the red head or the freckles that gave that impression. It was the glint, or, more accurately, the glow in those sunny eyes that did it.

Jimmy was three then. And now, approaching thirty, his behavior would indicate that the fire is still burning. Jim was up long before dawn this morning, as he is five mornings a week, driving a milk route so that he can earn what it takes to buy a beautiful old Ohio farm, and so that evenings and weekends he can work still more long hours to maintain it. He's the one who had the extra time and the energy it took to spark a full day of special activities for this spring's commencement at his alma mater, a little Quaker boarding school.

Of course, Jim attracts others. At times there's been quite a crew in the old farmhouse. Some have lent a hand. But mostly they've come to "sit by the fire," so to speak, to steal a little energy. When they get the message that stealing energy isn't quite cricket, that they should be putting out a little steam themselves, they move on. Jim's not the only guy around with energy. Of course not. Lots of people have energy. Lots of energy. But, where do they get it? What kind of energy is it?

I think of Bill and Peg. Boy, did they work! They ran a bookstore, one of the best in Southern California. They used to win prizes for their window displays. Those displays were good. I'd come to work on a morning and find they'd been up half the night re-doing things, from window layouts right on through the store all the way back to the storeroom, Bill puffing around in his underpants, trying to beat the heat of the effort. But there was something amiss, and I couldn't keep out of my mind Sam Johnson's sour remark about bookstore folks. "Vultures on the tree of knowledge," he called them.

Anyway, it wasn't the love of knowledge that kept them there, working so hard. Or even the love of window displays, though Peg was a real artist in that line.

I remember checking in early one morning. I always did check in early, so that is not the reason for my remembering the day. Rather, I sensed something was amiss, more than usual. Soon after they unlocked the door to let me in, and I got to my duties back in the store, an old friend of theirs came to chat, and they unlocked and let her in.

"Mary!" Peg was quick to begin. "Mary, the doctor has taken Bill and me off alcohol *and* gambling!" There was a long silence. "God, Mary, what's there left?"

She wasn't kidding.

She was dead serious.

It wasn't news to me that alcohol and gambling mattered to my employers. I knew they flew off every weekend to Las Vegas. If they were prompt about opening the doors to business every morning, they were even more prompt about closing them at the end of the work week for the business that mattered most. But what I had not ascertained or even imagined was just how much it mattered. The doctor, in a real way, had pronounced an end to life as they lived it.

It was over! All over!

I didn't stay long enough at the bookstore to learn whether Bill and Peg found a substitute reason for living. I don't know where they went after that for their energy. If they found new crutches, I wonder if they were as crippling as the old ones. Maybe, though, they just decided to live it up. Or, to put the same thing another way, maybe they decided to just blow it all.

"I eat what I jolly well please," says a neighbor-housewife. "If it gives me a headache, I just take some aspirin. I mean to enjoy myself, by golly." She gives her husband the same pleasures, three times daily, at least, and almost lost him this winter.

And now the "doctor" says that the environment is ailing. Feeding too much poisonous stuff into the air and streams and soil. We've been ordered off so much of this gas and coal, told to buy smaller cars, drive them less. So, what happens? Gas consumption goes up, big cars continue to sell. We mean to enjoy ourselves, by golly. We mean to have that energy.

You'd think we lived for the stuff. Maybe, in fact, we do.

Of course, some folks are a little smarter than the rest. They're going solar. Solar's the big thing these days. Just putting a solar panel on your roof to pre-heat your hot water wins you one full-size halo, and think what you can win by covering the whole roof.

Wings, almost.

By what's going on here? Is this the same old game? One wonders how many vultures can roost on the new solar tree.

This energy business may turn out to be a complicated thing. You dig in a bit, and what do you find? Roots. All kinds of roots, and they go everywhere.

Most often when we talk about energy, our attention is focused on things like oil, the Arabs, natural gas deregulation, the scrap over nuclear, how to get the sticky fingers of the utilities out of our pockets, how to get our own sticky fingers onto more bucks with which to pay for the energy and the products of energy we already consume, and, if possible, to buy still more of them.

What is going on here, really? Are we hooked on energy? Is that the problem?

Or is it that we're hooked on the wrong kinds of energy? If that were the case, we might give up petroleum and nuclear, and take to that good, clean, solar fuel. That would take care of the problem, set everything aright, wouldn't it?

Or would it? Somehow, the doubts keep rising, the questions persist. In fact, the doubts and the questions get more troublesome. Do we have an addict's relationship to energy and its use? There are aspects of our behavior that seem to suggest it. But, why would it be so?

Could it be we're addicted because we've gone after energy for the wrong reasons?

And then, the frightening question: Are we so addicted, so dependent, so crippled without our energy fixes that we would blow it all rather than give them up, blow it, because there's nothing else left? Energy is the stuff of life. Take Jim off his energy sources, whatever they are, and that project to save the fine old farm would simply collapse. It would shut down, just as surely as would a steel mill cut off from its coal supply. The hundreds who enjoyed Jim's very special Commencement Day would have spent, instead, one of the plain, ordinary kind.

It would be nice if we could point a finger at energy "out there," and swear never to touch that dirty, wicked stuff again. But, we can't.

One cannot help suspecting that the doctor's stern orders to Bill and Peg might have been treated as the occasion for re-examining a way of life that was not exactly cheerful, anyway, and that had built right into it an unnecessarily early exit time. And the neighbor-housewife might have taken a cue from her headaches, improved her life and her husband's, too, and added years to both their futures.

So, what about the international energy crisis? Is it a message, an order from The Doctor? Are we being pushed into a corner, so we will be compelled to think, to ask some questions about what we're doing, and why we're doing it? Is there a Grand Hope operating Somewhere, a hope that we'll work our way out of this corner into a saner, happier, healthier world, a world with a long, long future?

Whether in our personal or in our corporate lives, it is clear that we have not yet got hold of energy that is renewable, that is a boon and a blessing to our lives and to the total environment. So, we're on the substitutes. We keep ourselves going on greed, anger, hypochondria, bumming, Arab oil, alcohol, junk food, pot, big cars.

We're out of kilter. So, inevitably, we knock the environment out of kilter, too. The environment is sick because we're not so very well ourselves.

It does seem to be a crisis time for human beings, and for our societies. Gerald Heard, historian-philosopher, observed years ago that modern nation-states fall into two classes—the dictatorships and the democracies. The former are held together by fear; the latter are held together by greed. Aren't they equally unviable, in the long run? You can't make either of those glues hold indefinitely. In each case, the reasons for eventually coming unstuck are intrinsic, inseparable from the glue itself.

So, where should we get our energy? Do we go nuclear? Do we go solar?

Does it really matter? One has a growing and uncomfortable suspicion that it probably doesn't matter, if we keep on missing the real question that is being asked of us—Where does the *human being* get his energy? If our answer continues to be greed, or one of greed's close relations, we will probably do with solar pretty much what we have done with the fossil fuels. Oh, we may clean up the environment some, enough at least to get by. But we'll continue to have problems. We'll continue to teeter on the brink of blowing it all. We'll have one "doctor" or another, trying to wake us up, trying to get us to discover more than we understand now about ourselves, about the sources of human energy.

Though it probably can't be rated as more than a mere beginning, it's not a bad way to begin just to know, early on, that "I gotta FIRE inside a' me," and to manage one's first thirty years without putting it out.

Foonmanship

By Malcolm Wells

Felton J. Foon's ecologic dilemma:
What in the world should he plan to do first?
Clear off some land for a vegetable garden?
Dig a big rain-pit to ease his yard's thirst?
Plant a dense hedge to reduce the street noises?
"Order some storm sash right now," said his wife.

How can I choose when I have such wide choices?
Job number one may be: save wildlife.
"Natural, vine-covered, hideaway beauty.
Maybe that's first on our list," said his spouse.
"Then we'll put paper and garbage in compost;
Solar collectors all over the house."

Well, as you can imagine (having felt the same uncertainties yourself), it wasn't easy for the Foons to establish their ecological priorities. Everyone offered a different opinion. The repairman said, "Absolutely, you've got to convert right away to solar heating." "Nonsense," said the landscape architect, "make your place naturally beautiful; it'll inspire you to do all the rest." "Wrong," said Aunt Barbara, "Vegetable garden first: look at food prices. Think of all the starving people."

So the Foons did nothing. Absolutely nothing at all. The air over their town got a little smokier, the river a little smellier, the utility bills a little higher each month. The Foons' inaction was bad for the national economy, too.

Then, one day, Dooley Naze came along and told them about the wilderness graph.

"Sounds real exciting," said Foon, trying to act neighborly.

"Felton," said Dooley, "five minutes with this graph will do more for you than a year with Ralph Nader."

So the Foons took a chance, tried the graph, found their priorities, and started setting things right. That's how Foonville got all well again, and how the depression there was ended.

Before long, the Foonville saga had spread all over America. In no time, hundreds of towns were using the graphs. Then thousands. They brought the nation a prosperity no longer based on Arabs, nukes, and Fords.

Still, here and there, pockets of depression remain—places where the graphs are unknown. (If you happen to live in one, maybe you'd like to see how they work.)

All you do is compare your property to the perfect + 1500 score that it got, back in the days before Columbus arrived. Then you can see at a glance what your deficiencies must be.

Here's your neighbor's graph, for example. See how his rain and sun deficiencies stand out? He'd be silly to start correcting something else before he worked on them.

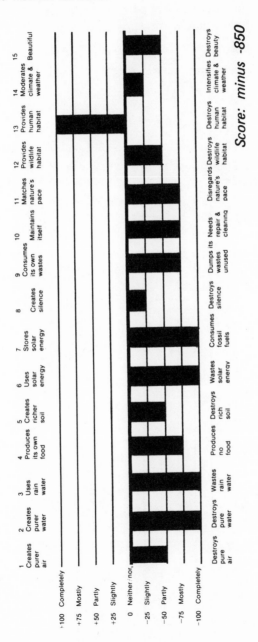

Now it's your turn. Just pick the values that seem correct, and see what they show.

If they show a score above minus −500 I'm going to put your name in for a congressional medal—or have the auditors take a closer look at your graph.

	1	2	3	4	5	6	7	8	9	10	11	12	13	14	15	
	Creates purer air	Creates purer water	Uses rain water	Produces its own food	Creates richer soil	Uses solar energy	Stores solar energy	Creates silence	Consumes its own wastes	Maintains itself	Matches nature's pace	Provides wildlife habitat	Provides human habitat	Moderates climate & weather	Beautiful	
100 Completely																
+75 Mostly																
+50 Partly																
+25 Slightly																
0 Neither/nor																
−25 Slightly																
−50 Partly																
−75 Mostly																
100 Completely																
	Destroys pure air	Destroys pure water	Wastes rain water	Produces no food	Destroys rich soil	Wastes solar energy	Consumes fossil fuels	Destroys silence	Dumps its wastes unused	Needs repair & cleaning	Disregards nature's pace	Destroys wildlife habitat	Destroys human habitat	Intensifies climate & weather	Destroys beauty	

Your score:

Drip. Splash. Creak. Groan.

By Malcolm Wells

For some reason, I'm fascinated by underground spaces. Mines, caves, bunkers, cellars, vaults, subways—they have great appeal to me. That's why I always try to get my clients to build into the earth.

I've designed twenty or thirty earth-covered buildings, and most of them have stood up quite well; only a few collapsed when the builders began to pile the dirt back on their roofs. But I think I've got that little flaw licked now. Now I fill every room with a maze of steel posts to prop up the structure. Now, when the dirt gets piled on top, the worst that happens is some cracking and sagging, and I can turn my attention to other little problems.

One of them, of course, is dampness. You can't go underground and expect to stay dry. But I *have* learned how to divert underground streams, or at least slow their currents so they no longer carry all the furniture away, and it's pleasant, down there in the total darkness, to hear the sounds of water flowing all around you. They cover up all the little creeping, scurrying noises. The only thing about the moisture that still bothers me is this endless dripping from all the overhead surfaces.

Another small annoyance is the constant mildew and mold. It has that awful smell and makes everything rot. Books, food, clothing—nothing lasts very long down there. But, fortunately, that's not important; it's so clammy-cold no one stays very long, anyway.

The only real problem is the gas. My son, John, suggested we simply give each person a canary, and we tried it for a while, but we were losing fifty or sixty a week. Dropping like flies. Now we just warn people and let them fend for themselves.

It's a shame, isn't it, that underground architecture has these nagging little problems? It's such a great way to build. That it costs only two or three times as much as above-ground buildings means it's not out of our reach, financially, either.

Slimy, damp, dark, and expensive—that's my kind of architecture.

Didn't surprise you a bit, did it? Exactly what you'd expected to hear about earth-covered buildings, wasn't it? I might have known.

Here I am, practically worn out from showing people how bright and pleasant underground buildings can be, and you're still picturing dungeons, even after reading all the articles and seeing all the TV shows about these sunny rooms. Even after thousands of people have visited them and said, "Look how bright and cheerful they are!", you're still ready to believe the worst. No wonder architects who build this way have so much trouble selling their ideas. No wonder they try to find so many other names for what they do: "terratecture," "geotecture," "earth-sheltered architecture," anything but that awful word "underground."

The truth is that underground buildings—at least the kind most architects propose—are naturally sunny and bright. They open from the sides of hills, or into sunken sun-gardens. Underground buildings do cost a little more than other buildings; after all, they have to carry the extra weight of those beautiful rooftop gardens. But they offer us a healthy alternative to our dying cities of asphalt and concrete. Imagine: rooftop wild gardens downtown!

Underground buildings need very little outside maintenance, and they offer excellent fire protection and storm-proofing. They're unbelievably quiet, and—best of all—they save so much money in heating and cooling bills that their extra costs get repaid often in less than ten years.

I could go on all day, in fact, but there's really only one big reason to build this way. It's the very basis of life on our particular planet. Biologists tell us the life-scheme here is this: green plants covering the *entire surface* of this sun-bathed sphere, every inch that's not ice, desert, or rock. Our lives depend on it.

So why in the world do we build all our highways and buildings in the only way that's sure to destroy us?

Looking for a New Aquaculture: The Roads Ahead

By John H. Todd

If you're near Cape Cod some Saturday, go see what the new Alchemists are doing there near Woods Hole. Amazing. Bioshelters, wind machines, solar fish ponds, and some of the lushest vegetable gardens you ever saw. The "Alchies" are a varied group, and John Todd by no means gets all the credit for their success, but he's the driving force. And he no sooner gets the New Alchemy Institute going than he's off organizing solar villages, and then Ocean Arks—sail-driven cargo ships—floating greenhouses. He lives a generation or so out in front of us.

To say that my life changed that day would be an oversimplification.

For a long time, I had wanted to have a look at a renowned experimental fish culture installation—the brainchild of a distinguished scientist. If I told you I had been invited to visit, I am afraid I would be stretching the truth. My crony and I were much younger then, and we tended not to stand too much on formality—which is a windy way of saying we decided to sneak in.

One morning, as casually as possible, we sauntered in, poked around, sniffed the air a bit, and inspected the model facility, which was touted in all the best papers as the aquaculture farm of the future. My friend, Bill McLarney, and I were filled with high expectation. We earnestly desired to assess the state of the art. And being neophytes to fish culture, this kind of visit seemed the only way to go about it.

As it was early in the morning and no one was about, we looked around as best we could. I was impressed. No, that's not accurate. I need stronger language. Maybe awed is the word. Bevies of pools, batteries of pumps, aerators, filters, back flushers, ozonators, infrared irradiators, filters, timers, feeders, and miles of wiring and plumbing greeted our eyes. Then there were the sensors, probes, records, and electronic gear that took our breath away. We were seeing space-age aquaculture firsthand.

The whole rig looked like a NASA (National Aeronautics and Space Administration) space lab with the cowlings off. I can't tell you how excited I was. I was sure this demonstration of an engineering triumph should sustain undreamed-of numbers of delectable creatures.

Stepping and stumbling over the paraphernalia, we inched up to the pools themselves. Whipping out our Polaroids, we prepared for the real show. Containing our anticipation, we peered into the swirling water.

At first, I saw nothing. Then, as my Bausch and Lombs—and my eyes—began to acclimate,the life within revealed itself.

"Hey, Bill," I said, "There are fish in here."

I started to count the fish—using my usual one finger at a time method so the numbers wouldn't get away from me. The fish were swimming by slowly. By the time I had worked my way up to the fourth finger on my left hand, I had exhausted the population. Not believing my eyes, I waited until they came around again. This time, I managed to count my way to my thumb.

"I guess you missed one the first time around—or squeezed two fingers on the second counting," snickered Bill.

"At this rate," he went on, "it looks as though there are more electronic pumps than fishes around here. It must be one of those utility-sponsored projects."

It would be an exaggeration to say that this is all we saw. If memory serves me correctly, there were, in one or two of the pools, a few milk-carton-sized racks filled with some clam species or other—perhaps a pioneering polyculture experiment was being launched.

We repaired to a local saloon to calm down a bit. Later we learned from official sources that what we had seen was a prototype. It still had bugs in it. But bigger things were in the works, and solid government backing was believed to be just down the road. It seemed to our still-innocent eyes that this type of techno-twit's vision might engineer aquaculture out of existence in this country. All that paraphernalia seemed a blatant denial of nature's skills.

I wondered if the ex-aerospace guys trying to manage us with their toys know that the living world has equivalents to their expensive machines—equivalents that are cheap, self-repairing, reliable, and don't need fossil fuels. About the only thing the fancy hardware can do that nature can't is make millionaires out of manufacturers and engineers.

Don't get me wrong—please. I am no stone-ager, advocating technology-free fish farms. There are times when we mortals need help. Besides, anybody who has seen the wind-engines, bioshelters, and solar aquaculture courtyards at New Alchemy knows that we are into technology—but the kind that serves natural or living systems—and only as a junior partner.

Since that day, I have done much reflecting. I am coming to believe that many of my fellow aquatic scientists are working from an incorrect premise. Namely, they are trying through various ingenious means to make aquaculture environments match precisely the needs of a few commonly cultured fishes, such as trout. I have listened with patience to their arguments for doing so, which revolve around market potential, consumer tastes, and sales prices per pound. Their considerable skills are focused on vaccines, medicines, nutritional needs, and chemical and engineering components of culture systems. Yet the best efforts of research and development programs scarcely seem to alleviate the energy, capital, food, water quality, and disease problems that plague aquaculture.

What is the answer?

There is no single answer to anything that's valuable. There is a road that can be followed, and I want to explore it with you. After that fateful day, Bill and I began to think in terms of inverting the whole equation for aquaculture—using as guides modern ecology and the collected experience of oriental fish cultures.

We decided to begin by designing and assembling productive environments and stable ecosystems within which fishes could be grown easily and cheaply. That was Stage 1. During Stage 2 we set out to find the fishes to fit the designed environments. The whole globe became our beat.

This approach was not quixotic. In less than a decade, several of the species we have cultured in semiclosed warm-water ecosystems have developed real market potential. The best known, tilapia or St. Peter's fish, sell in large cities, including New York and Chicago, at prices that would swell the heart of any fish farmer.

The long-range strength behind this ecological-design and fish-matching concept lies in the fact that there are so many fishes to choose from. Most of us really aren't aware of this. I would be willing to bet that fish will provide most of the new foods for our hungry world over the next hundred years. Fish are the last, real, untapped food resource for humans. Despite having had a globe-wandering fish gourmet for a doctoral professor, it has taken me a long time to overcome my deeply ingrained prejudice that trout, salmon, and catfish are the only species worth bothering with.

There are probably more fish species than all other vertebrates combined. If one takes all the birds, frogs, lizards, and reptiles, including snakes and turtles, and adds to these all the mammals from mice to elephants, blue whales to people, the total number of species approaches 21,000. Estimates of modern fishes vary from 15,000 to 40,000 species. It may well be that, when and if they are all discovered and identified, their actual numbers will settle out at approximately 25,000 species. A goodly percentage of these reside in fresh or brackish waters.

We don't know much about all these fish species; only a small fraction have been cultured. Beyond this, very few have been evaluated as candidates for fish culture. I applaud the Fish Culture Section of the American Fisheries Society for their addressing this problem at Aqua-culture/Atlanta/78.

Yet, with some temerity, I would suggest that aquaculture is suffering from the lack of a wholistic, ecological perspective. Most polyculturists use a handful of species culled haphazardly from world stocks. We don't know that freshwater prawns, tilapia, or Chinese carps are the way to go. Other species may have more to offer.

One of the great, untapped reservoirs for new human foods is South America. The Amazon and Orinoco River basins gave birth to over 2,000 species of freshwater fishes. Yet, this continent, which has the richest and most diverse fish fauna, has scarcely been explored by aquaculturists. The range of fish types is enormous. One—an armored catfish—is a veritable mountain climber. It has a holdfast organ, which is part of its mouth, that it uses to scale mountain streams to an altitude of 15,000 feet. More endearing in a culinary sense are species of characin, South American catfish and cichlid families with a reputation for being pure ambrosia to eat.

Nobody knows how many of these fishes might be cultured in ecologically derived environments. There are some clues to support my case. One species we have worked with in Costa Rica is a characin, *Brycon guatemalensis*—known to the local people as machaca. The diet of this fish would gladden any fish farmer—it consumes terrestrial detritus like leaves, wild fruits, and nuts, almost anything that falls into their infertile stream homes. Their teeth would make a shredder-grinder executive green with envy. Machaca used the biologically inherited machinery in its mouth as potential replacements for shredders, choppers, and spreaders down on the fish farm, while growing fat and tasty on what, in our society, are normally considered wastes.

I am not suggesting that the machaca is the only answer to our problems. We have all heard that story before. I do think, though, that unknown riches await a systematic search of the fishes of South and Central America.

My tale has a tragic shadow side. The search for fish species suitable for culture must begin soon, because the New World tropics are being deforested. Streams are silting in as the surrounding forests are devasted. Many fishes are threatened. Extinction might be in store for some. How quickly the ecological destruction is proceeding can only be guessed. But, when one man—America's only multibillionaire, Daniel K. Ludwig—has caused to be cut down millions of acres of Amazon jungle, equivalent in area to the size of Connecticut, to grow a single species of tree for pulp and paper, you can bet that the fishes indigenous to this private fiefdom will suffer.

I seem to be announcing that there are untold new fishes available for culture and, at the same time, slamming the door—saying they might not still be there for much longer. If it were to become widely known that South American waters contained biological "gold," the ecological destruction of the forests might be slowed. Cynically, I suspect the possibility of wealth to be the basis of decision for many people. By proving that wealth lies not in exploitation but in the careful husbanding of aquatic environments and in the forests that surround them, then the fishes stand some hope of being saved and, in turn, might provide a foundation for New World aquaculture.

I should clarify what I mean by the use of these fish in aquaculture. I am not advocating that Amazon or Orinoco River fishes be stocked in

North American lakes. The casual introduction of exotics can prove disastrous. I am recommending rather that large numbers of them be studied for their culture potential. The most promising could be cultured in their native areas, in tropical areas where native fishes are lacking, and in solar-heated, semiclosed culture ecosystems in northern industrial countries à la New Alchemy.

Having covered so much ground, I should like to retrace my steps briefly. I have said that aquaculture is headed down the same cost-spiralling road that threatens agriculture. The addiction to fuels, chemicals, increasing amounts of machinery, and rising land taxes, etc., which cause production costs to skyrocket at a rate greater than the market value of foods at the farm gate, should cause us to reconsider our general direction. With this scenario the last person to make money is the fish farmer.

I am proposing that aquaculture diversify its options. This has been said before. I am merely urging that aquaculture shift toward a more complete biological model. It should shift the emphasis from engineering-culture environments that precisely fit the needs of a few fish species with narrow and specific environmental and water-quality requirements to the designing of environments that are ecologically balanced, stable, and productive.

The name of the new game is the substitution, where possible, of ecological knowledge for capital, energy, and hardware. The next step would involve the enlarging of the fish repertoire of the aquaculturist by finding species that match biologically designed environments. Finally, I would suggest that—in the task of building an aquaculture that can thrive in a time of economic and resource uncertainties—the fishes of the world have scarcely been tapped.

I would be very interested in hearing from readers who are evaluating various fish or who have leads on potential species for culture.

(John Todd's address is: The New Alchemy Institute East, P.O. Box 432, Woods Hole, Massachusetts 02543)

"Turn A Light On; You'll Ruin Your Eyes!"

By Malcolm Wells

How many times have you heard that? If you're at all like me you've been hearing it—or saying it—all your life.

Those same words were spoken by Abraham Lincoln's mother almost a century and a half ago. She was watching Walter Cronkite one evening, and, during a commerical break, she noticed that young Abe was back to his old tricks again, chalking arithmetic problems on a shovel by the light of the fire.

"Turn a light on; you'll ruin your eyes!"

History does not record the future president's response to his mother's warning, but since the incandescent lamp was not to be invented for another 50 years, it seems safe to assume that the boy did not turn a light on; he probably quoted the great optical truth we're just rediscovering today:

Insufficient light may cause eye fatigue but not eye damage.

Next time you go to the supermarket, or visit a schoolroom, take a look at all that unnecessary light—at all that oppressive glare—and realize you're seeing a dinosaur, a relic of the years when energy was cheap, and power companies said dim light hurt eyes.

Every few years, newer and higher brightness standards were established for every human activity. (Well, *almost* every human activity.) Sight is so precious, and the standards sounded so authoritative, we all ran out and bought bigger light bulbs and more lighting fixtures until you couldn't enter a library, a drugstore, an office, or even some houses, without reaching for your sunglasses.

Then, along came the energy crisis and up went our electric bills. Now the power companies are so busy cashing checks they simply don't have time to tell us their lighting standards were too high. Most rooms are as bright as ever. But don't let it worry you; with all the new emphasis on nuclear power and new oil exploration, we should soon have power to burn (if you don't mind genetic damage and lung disease).

It was at a recent conference on buildings and the environment that I first heard about lighting standards being largely nonsense. One of the experts there tore the eye-damage hoax to shreds, and, surprisingly, even the power company people in attendance took no exception to what he said. To me, it was the most important news of the day. It had that nice, logical ring to it that so many simple truths have. I'd always squinted in supermarkets, and I'd often found reading at dusk to be pleasant, but I didn't want to depend on such subjective evidence, so, the first chance I got, I made an appointment with a highly respected opthalmologist, and asked his opinion.

He assured me that, although insufficient light may tire the eyes, it will not injure them. He said, in fact, that a great deal of eye-fatigue comes from too much light, and that fluorescent tubes often cause an annoying flicker at the periphery of one's vision—out of the corner of your eye, so to speak. But I'm not panning fluorescent lights; far from it. They're many times more efficient than the incandescent kind. The main point is that we need no longer bathe entire rooms with such glare, nor do we need as much light on work surfaces as was formerly assumed. The present trend toward energy-conserving "task lighting" not only saves oil and coal (and land and lungs and oceans), it can also reduce eye-strain.

There's something very pleasant about moving closer to a lamp when close work is required. (I'd have to move very close indeed if I wanted to thread a needle.) It's nice to move from a darker area to a lighter one and back into the softness of the shadows again. Rooms look more interesting with pools of light, and the poetic quality of nighttime is more apparent.

The poetic quality of lower electric bills is more apparent, too. That's the nice part about something like this: you gain in every way. I think we're going to find a richer life on the other side of this energy crunch.

Now if we can just manage to live through the economic side of it!

Energy in a Bucket

By Malcolm Wells

This message is actually being written inside the White House, without the knowledge of anyone. It's about cover-ups. A recent public opinion poll showed that although 67 percent of those queried didn't understand the questions, and 74 percent had no opinion, a whopping 89 percent admitted to having actually participated in cover-ups of their own, and therefore felt they could not criticize others who, in the interests of national security, had done the same.

National security has, in the past, been stretched to include all kinds of activities, but the energy crisis has seldom been recognized as one of them. When you think about it, this is a strange situation because the energy crisis is very much a national security matter.

Suppose all the lights went out just before the Chinese hordes entered Seattle. Then where would we be? Or if the fuel shortage caused so much warmth-cuddling that the population boomed again and we had suburbs from coast to coast and the Air Force had no place to land? Or if we all ran out of gas and filled the streets with bicycles and there was no space left for a military parade? Aren't these questions reason enough to classify the energy crisis as a national security matter? Of course they are, and there's always the larger issue of draining away the resources needed for vital things like electric toothbrushes and neon signs.

One of the easiest ways to stop wasting energy comes in buckets. It's called white paint, which, when it is applied to darker walls, makes rooms so bright you need only half the light bulbs you formerly used.

That's why I just painted all my walls white. And it's why I'm now in a white house.

Windows can make a room brighter, too, of course, but they lose far, far more energy in the form of heat than they gain in the form of light. The day of the unprotected window—even of the double glass kind—seems to be drawing very rapidly to a close.

Pine panelling, natural wood, brick interiors, stone walls, barn boards, rich colors, dark floors . . . all these are now no-no's. Shine a lamp on polished wood and what do you get in return? Luster. Shine a lamp on a clean, white wall and what do you get? Light! Like the oyster and the clam, we must learn to use rich, dark, earthy colors on the outsides of our houses, and smooth, clean, whiter colors within.

You can actually reduce the lighting part of your electric bill by more than half if you make your walls reflective. Your paint dealer can tell you the reflectivity of every paint in his shop, so you can't go wrong. You'll save money, you'll conserve natural resources, and you'll lessen the threats of more strip mining, more oil waste, and more nuclear power. So if white paint isn't a cover-up in the interest of national security, I don't know what is.

Conservation Is a Computer Solution

By Jeffrey Cook

"Don't get confused by the facts," says this professor of architecture at Arizona State University. "I am overactive, overworked, and undersupported." Still, he finds time to be on the boards of the big solar organizations, to organize solar conferences, and to turn out an impressive number of books—on Bruce Goff, on cool houses for desert cities, and on passive solar architecture. Meet Jeffrey Cook.

Conservation is a twelve-letter word. It is an accretive word with at least as many meanings as syllables. As a word it is neither lean nor clean.

Conservation is not a happy word. It connotes doing less. It implies miserliness. Its Scrooge-like images suggest a regression of standards—minimums in both inputs and outputs. It is associated with those occasional ascetics found everywhere who prescribe frugality and even perpetual sacrifice. But they are not unique to our time.

For misplaced Calvanists, conservation may mean suffering a bit more in this life to be more worthy of the next. For the self-righteous, it may be an excuse to challenge inept local utilities and faceless multinational corporations. But social indignation and gross human exploitation should not need resource scarcity to galvanize action.

In fact, the reality of many personal sacrifices in the name of conservation may simply be the saving of precious resources for the unappreciative squandering by one's neighbor.

Thus, the task of convincing the world that conservation is the salvation of humankind is a depressing challenge. Conservation may become the word for lifestyles dampened by economic restriction or oppressive inflation, but it is unlikely to be the choice of those who have tasted the good life of generosity and continue to have a choice. Potentially, conservation may become the unchosen standard only for those sufficiently unappreciated by society to be allowed no choice.

Of course, there are many businesses that have found that conservation is an easy way to make a buck. But saving money to make money is not a piece of higher consciousness. In contrast there have been highly regarded cultures, such as medieval Japan, that have been based on a design discipline of limited resources. But those have been ethical and even religious undertakings, which have been based on well-defined hinterlands of modest resources. Such geographic determinism is not the case for a wealthy half-continent or world wheeler-dealer like the United States.

The voluntary assumption of burlap when double knits are the social standard seems unlikely. Shortened showers of undependable temperature may stimulate the complexion and dull sexual drives, but they won't win either smiles or votes. Red, white, and blue chills won't become the patriotic honor of a country of gregarious do-gooders and well-wishers. In the name of the planet, not many Americans are ready to turn down the thermostat. Neither are they yet prepared to take their collective foot off the accelerator, whether of their automobile, their country, or their several varieties of space ships. Americans are expecting more from life every day, and "more" means quantity.

Conservation is obviously an ethical issue. But it is an ethic that most people would prefer someone else to practice. Partly, conservation has to do with holistic perceptions—the understanding of the nature of the world and the relationship of man to all its resources. Understanding has always been a scarce commodity. But there is hardly a basis for holistic consciousness when many people do not even know where the hot water in their own home comes from.

Conservation as a compulsion of giving up something is a negative obligation. But turning off the lights so one can use candles is not a repressive act—it is a rearrangement of values—and one which, incidentally, may not save the world. But it does represent choice and consciousness. It is also an act that enhances life.

Knowledge and consciousness are conditions preferred in most areas of human concern. But consider two recommended household conservation acts: adding insulation and turning down the thermostat. Even from a selfish point of view, one will potentially increase comfort the year around; the other will provide discomfort. Both are inert acts—their meaning and value depend on context. Conservation as a word does not communicate the comparative value or purpose of those acts.

Yet, the joys of being generous about the daily events of life have some poetic potentials, even within a future of scarcities. Why not leave the windows wide open on a chill, fall day to smell the tang of burning leaves and feel the nip of oncoming winter—if one's space is freely heated by the sun? Somehow, clean dishes sparkle more with too little detergent and with too much solar hot water. Homebaked bread fresh from the solar oven in the snow-filled yard on the clear afternoon after a snowstorm is a fusion of many natural cycles. Perhaps the cosmic rhythms inspired by *Star Wars* fantasies have a higher plane of consciousness when the "force" is the ultimate energy generator. The sun as daily supplier of both needs and pleasures means going to the source, and not a secondhand adventure via an abstract medium with all those attendant pollution and distribution problems. It also implies a new exuberance in things physical. Such a lifestyle would not include conservation in any conventional sense.

Quality of life has less to do with the package than how it arrived. Turning the switch to "on" gives total delivery control but indifferent meaning. Human enthusiasms are often based on the attitude of participation, and only partially on the perfection of performance. Few beautifully regulated electric meters have been immortalized as one of the arts. But all of the extant solar air thermosiphon houses have received repeated literary applause; one with blatantly unscientific, but high aesthetic, sensitivity.

The recent invention of "energy-conscious" design is only a footnote to the sudden, shaky devotion to conservation. Design has always been energy sensitive. Design has always been a discipline of balancing resource values against promised futures. The triumph of the dome over St. Peter's was a deliberate value statement of priorities against the dilapidated housing of the common man in its shadow.

With the fright of disappearing fuels, the scramble for new ethics has been a reality. But energy issues have always been critical. Priorities have always been made. Design has been operative even when apparently absent, because decisions are always made. Thus, the sudden discovery of an "energy-conscious" design tradition has been an easy and transparent ploy.

The illusion that "energy-conscious" design would be conservative is naive. Consciousness must remain consciousness. Some forms of consciousness require exotic materials and extravagant quantities. Others don't. How many tadpole eggs produced those few old croakers that harmonize September twilights? Other events in nature, such as eclipses, are rare. Nature has no consciousness—either of preserving or of squandering. Nature is not a numbers game. And human life may not be either humane or lively with rations and quotas. Conservation is a computer solution for a question of human values.

The Topsoil Wars

By Malcolm Wells

Although they were viewed by many historians as but two parts of a single militaristic phenomenon, the fuel wars of the Eighties and the topsoil wars of the Nineties were in fact separate events; Arab against Jew in one, Arab and Jew allied in the other. And, while the unprecedented slaughter during the fuel wars was somewhat mitigated, later on, by the respect it generated for man's energy-needs, the topsoil wars have done nothing but push us to the brink of anarchy and starvation.

No single cause of the topsoil wars has ever been identified, but it seems not unreasonable to say that the frenzy began soon after the presentation, at The 1987 International Scientific Congress, of the paper, "Human Needs." It was then that topsoil was first identified as by far the most precious and also the most endangered resource on earth. Air and water manage to rid themselves of most impurities within months or years, but topsoil, once destroyed, takes thousands of years to be replaced. The report, which carried the endorsements of not only leading scientists but also historians and statesmen, proved that great civilizations had flourished only in areas of high land-fertility, and that in spite of modern technological advances, civilizations would continue to flourish only when they rested upon a solid base of rich, deep soil. Also in the report were disturbing indications that many human diseases were less prevalent among people whose food was grown on naturally fertile land, free of chemical fertilizers and pesticides.

The very next month, prices of organic topsoil began their climb. Coupled with the artificially high prices of food, the soil-price rise started a worldwide search for loam. By 1990 it was quite evident that in spite of the conversions of many oil-carrying ships to soil-transporters, the clamor for untainted soil would never be satisfied. The sheer bulk of the material stood in the way.

In the summer of that year, when the topsoil-rich nations cut off all further soil exports, hijackings began to be commonplace, shipments of counterfeit soil caused a panic, and prices went through the roof.

The immediate effect upon the average property owner, wherever in the world he happened to live, was perhaps a good one. He tended to conserve and enrich whatever land he had. Here in America, where we had for so long been out of touch with the entire sun/soil/food relationship, organic gardening was taken up with a vengeance. Even the U.S. Department of Agriculture and its retinue of giant agribusinesses tried to restore the natural fertility that their short-sightedness had bled from the farms of America. But as massive soil-thefts increasingly threatened the stability of rural communities, and entrepreneurs began to hoard great piles of topsoil, dollars became nearly worthless in the scramble for food. Prices of topsoil on the European markets hit new records daily, finally surpassing the price of gold itself. Soon after that, the wars started.

Memories are too fresh to need reminders of the toll taken by this latest world-madness. That nuclear weapons went unused for fear of further contaminating the land was at least one blessing. It saved the few good soils left on earth. The rest had long since been washed into the sea by centuries of careless farming, or had been badly poisoned by the misuses of chemicals, or had been built over by huge urban complexes. That's why all the battles were fought over the most remote farmlands on earth.

Now it appears that we have a kind of peace. The times could even be described as moderately hopeful ones. Our energy-needs have been assuaged by sunpower, our population halved, and our respect for the life-giving land is higher than it has perhaps ever been. Such factors could even be read as harbingers of a new and more enlightened civilization. But today's news brings a chilling announcement from the President's Canadian Pentagon: a massive new space venture is being planned. Its mission: to bring the moon's untainted soil to North America before the Communist world can get it. This can do nothing but inflame fresh wounds and lead us back toward war.

Some Energy Options

By Ray Sterling

During his first year or two as the director of the Underground Space Center at the University of Minnesota, Ray Sterling coordinated a study for the Minnesota Energy Agency that resulted in the bestselling Earth Sheltered Housing Design. *I see and am more impressed by him every time I hit the earth-sheltered-building conference circuit. He knows what he's talking about.*

Energy crisis, economic depression. These aren't very pleasant thoughts; in fact, with the affluence of most Americans today, they seem like only dim possibilities. Is the refusal of most Americans to believe that there will be an energy shortage similar to the myopia of Europe before the Second World War or the moral decay of the Roman Empire before its fall? On the one hand, there are the "technology-will-solve-all" prophets who point out the unimagined technological advances that have occurred in this century. Both have plausible arguments, and either with only a few adjustments in future events could prove to be right. This leaves the average citizen somewhere in the middle—not knowing whom to believe, knowing whom he prefers to believe, and usually predicating his actions, or inaction, accordingly.

Is it prudent today for the average citizen to take steps to be able to supply his basic needs in the advent of energy-supply problems and the accompanying effects that these would have? If he did, would he end up like a member of one of those small groups that have misread the cosmic signs while the rest of the population snickers?

Of course, the first energy shortages aren't likely to represent a sudden cessation of our energy supplies and the plunging of the country into eternal darkness. The first taste of such a shortage was during the 1974 oil embargo. The effects of the shortage were very different from region to region according to the energy supply-and-demand picture, but the link of energy and economics was made clear. Some large industries were forced to shut down to protect residential heating and essential services. There were many ramifications ranging from the inconvenience of long gasoline lines to the more serious loss paycheck from a layoff.

What are the possibilities of our energy picture in terms of the severity of the problems we may face in the near future and in the long term? Some general observations about energy cost and availability can be summarized as follows:

1. Oil supply is limited by political constraints. For example, oil production rates can be adjusted (particularly in the Middle East) to meet demands in the near future with rapid depletion of reserves or to allow shortages in the short term to occur but ensure a continued supply of oil well into the next century.
2. Cost of energy is also very political, at present, rather than a reflection of the true cost of production.
3. Energy cost will inevitably rise much faster than the general rate of inflation if prices are not held at an artificially low level.
4. If energy costs are held to a low level for political reasons, then shortfalls will occur much earlier than otherwise would be the case (unless some form of rationing is introduced).
5. The date at which energy shortfalls will start to be felt in the United States as a whole is generally considered to be between 1985 and 1995, based on present predictions of supply and consumption figures.
6. Local shortages may be felt earlier than a national shortage.
7. Depending on the effect of price increases or rationing on consumption, uncontrolled shortages may not occur in the near future.

8. As the price of energy rises, many sources of energy that have not been economic before can be added to the energy reserves picture. These reserves will, however, take a long time to be available to the consumer, they will cost more, and they will often have major environmental objections.

9. The economic system and the balance of payments of the United States will continue to be harder and harder pressed by rising energy prices and imports.

10. From a political standpoint, residential customers will be likely to be protected as much as possible both from crippling energy payments and cut-offs. This will throw a greater burden on the rest of our economic system.

The implications of this picture for the average citizen are that, while there is no clear certainty that there will be energy shortages of major crisis proportions during this century, there is a very good likelihood that energy prices will rise steeply or that some form of rationing will have to be employed to prevent unacceptable shortfalls. As our energy consumption and supply capabilities come closer together, there is more likelihood of local temporary shortages during times of peak use, which is also unfortunately when they are needed most. The energy picture will throw an extra strain on the economy and will take a larger and larger bite out of everyone's income. Jobs will be less secure.

The long-term energy future is obviously even more difficult to assess; it will probably depend more on whether the exponential growth in energy use, worldwide, can be slowed rather than on the extra conventional reserves that may be discovered in the meantime. We can switch much of our energy use to renewable resources, given time and the financial incentive of more expensive energy. If the consumption increase isn't slowed and we don't use significant amounts of renewable sources of energy, we will be facing immense problems in the next century.

I believe that the potential of serious energy problems within a relatively short period of time means that sensible citizens should examine their own lifestyles and shelter-projections for the future, and decide whether they will be in a position to:

1. Hold utility costs on their shelter to a reasonable level.
2. Survive supply interruptions without fear of "survival" conditions.

There are several ways to achieve these ends to a greater or lesser degree, and they are listed below, together with my own impressions of their suitability for the task:

1. *Add extra insulation, weatherstripping, etc., on existing houses*
   This is very good for energy conservation now, but for larger houses may not be sufficient to keep utility costs low enough in the future. These houses will not survive energy-supply interruptions in severely cold climates.

2. *Build modern, well-insulated houses*
   Again, this may not be good enough if energy costs rise steeply. Also, this type of house will not stay warm when the energy fails because of the lack of mass in the house.

3. *Retrofit active solar systems on existing houses*
   These will not be easily retrofitted to many older houses. They will have a high initial cost, require maintenance, and have a limited life span (20 years is often assumed). Nevertheless, as energy costs rise, active solar systems will become economically competitive in reducing heating bills. Without much larger storage than is customary enough for approximately three sunless days, the energy survival picture for these houses is only somewhat improved. The systems need electricity or standby generators in order to operate.

4. *Build multiple-unit dwellings*
   These do offer significantly lower energy consumption because of compact layouts and shared walls. An individual is, however, very dependent on the functioning of the building's energy-delivery system. Multiple dwellings, except those of the townhouse type, involve some basic differences in lifestyle from that in an individual residence.

5. *Heat with renewable resources*
   Renewable resources such as wood can provide an elimination of heating costs from other sources as well as long-term security against other fuel-supply interruptions. This option will not be available to everyone and will be costly if the wood must be bought commercially and transported over great distances. There will also be a potential problem of air pollution from large numbers of wood-burning furnaces in urban areas. Wood can provide a good backup for other systems, and the furnace's ability to provide heat without electricity should be considered.

6. *Build high-efficiency houses using active and/or passive solar energy*
These houses will have very low energy consumption when properly designed, and they will usually have larger thermal storage than that incorporated in existing houses. Passive solar houses have an advantage in that they can still utilize the sun's warmth when electrical power is interrupted. Most systems will not remain very tenable in a long period of supply interruption if the weather conditions aren't favorable. That houses are mostly custom-designed at present, and that the systems are not in common use yet, can lead to higher initial costs.

7. *Build earth-sheltered houses*
The possibilities exist for extremely low energy consumption in earth-sheltered houses. These houses can be combined with active or passive solar techniques to provide a low demand and large built-in storage of heat in the earth around the house. Due to the moderation of outdoor air temperature variations by the earth, the structure will remain tenable in long periods without normal energy supplies. Higher initial costs can again result from the lack of widespread experience and the custom approach to design. Costs are comparable, however, with other energy-efficient forms of housing and could be competitive with conventional well-insulated houses on larger multiple-house projects.

These options have been presented briefly on various ways to prepare reasonably for (and to help avoid) some of the energy problems we may face. More information on the advantages and disadvantages of each system is available in other literature. An encouraging point to note is that keeping energy needs to a minimum in our basic shelter does not have to result in lifestyle sacrifices.

Necessity has been labelled the mother of invention, and indeed there is ample evidence to support the fact that energy systems are improving. The important consideration, however, is whether these steady improvements in conservation will be adequate, and in time, because in a crisis everyone may be too busy coping to be very inventive.

Energy Economics

By Richard Livingstone

Solar energy's business expert, Dick Livingstone writes a column on the subject for Solar Age *Magazine, and he publishes a highly regarded directory of the solar industry.*

The alternative energy movement is not only alive and well but also flourishing. At least this is my impression after visits during the course of my work in energy education to dozens of New England communities—from its cities and suburbs to its rural villages—mostly in the eastern half of the region.

At the grass-roots level, meaning at the level of the "man in the street," that movement takes many forms and in fact varies from one community to another. There are working-class towns where the citizenry, hard strapped by rising fuel bills, is more interested in practical do-it-yourself energy projects, and towns where a professional class is more concerned with finding the right architect to design a passive-solar system. But the principal motive, varying in these towns only by degrees, is not the "energy crisis," with which there seems no great mass concern, but simply a desire to cut fuel bills and save money. At the same time, it also seems apparent that the ranks of the long-established core group of those with a conservation ethic, along with the energy tinkerers and experimenters in wind, water, and solar energy, is growing rapidly. Whatever the motive, one finds in these alternative-energy workers and users many of the old-time Yankee virtues, perhaps more prevalent in New England, of thrift, independence, and ingenuity, along with more modern motivations like a desire to beat the system— which probably accounts in part, along with a feeling of being "had," for the rage in New England, as throughout the country, among those who have cut back on their energy consumption only to find their electric bills still the same or even continuing to go up.

If there is a mass movement now to conserve energy, it is in the swelling ranks of those who have turned to or are turning to wood-burning stoves. This is a movement that seems to cut across all economic strata, from the family in the $100,000 home down to those in a modest $25,000 ranch home. It is not confined either to the country or the rural or suburban areas; dwellers in such New England cities as Lowell, Massachusetts, or Manchester, New Hampshire, are buying and installing wood-burning stoves in almost the same numbers as their country counterparts. That this is a mass movement the figures leave no doubt; a Strafford County Extension Service survey found over 50 percent of the households in New Hampshire using wood stoves as a sole or supplementary source of heat, and the study noted this may be a conservative figure, with other studies showing a percentage as high as 60 or 65 percent. The same high percentage (50 percent or more) has been reported for western Massachusetts; for Vermont the percentage has been put as high as 75 percent. And the wood-stove sales boom is growing despite a sharp rise in the price of firewood (now $100 a cord or more in the winter in such areas as Cape Cod and from $50 to $60 a cord out of season in New Hampshire) and a growing inability to obtain free wood (one person we talked to near Manchester had gone from burning oil to wood and then to coal).

Very few we talked to use their fireplace anymore, and this once much-touted appurtenance of new homes seems in the process of becoming obsolete, if it has not already (more new homes are coming equipped with wood-burning stoves). Another consideration (besides price) when buying firewood burners is that of the merits of various airtight stoves, whether you can get parts for a foreign-made stove as easily as an American-made one, that sort of thing.

If the turn to wood stoves has taken on the form of a mass energy movement, the growing army of energy experimenters must rank as the most significant minority movement. They are out there in growing numbers, at least in Maine, New Hampshire, and Massachusetts, some putting up wind-energy systems (often with their solar homes), others working on ways to harness streams and rivers on their land, and still others, perhaps in even larger numbers, building their own solar greenhouses. What is perhaps most striking about these individual efforts is that no two systems are alike; the ingenuity of New Englanders when it comes to dreaming up ways to tap the energy of wind, water, or sun seems to know no bounds; everyone seems to have his own answer to the best way to do it. Scratch an inventor or a mechanically minded New Englander these days and you'll find an energy experimenter and sometimes a budding solar entrepreneur with a new way to put the sun, wind, or water to work; one New Hamshireite, for example, told me of an idea he had for a home-made solar heater to warm the waters of his swimming pool, and a few weeks later, 100 miles away in Massachusetts, I met another who had put into operation a similar idea, alike in its essentials but unlike in its details. You feel that this is how it must have been in the early days of the automobile, when everyone was his own backyard mechanic and new automobile designs flowered from the hands of hundreds of thousands of independent experimenters.

Public interest in solar energy seems higher than it has ever been, just below sex, to judge from requests for information about it in libraries (at one library I visited in a small rural town in southeastern Massachusetts, three people in the space of half an hour came in to ask the librarian for information on solar energy). The interest is especially intense among school children, students in high schools down to youngsters in elementary schools; I was told of one third grader who had built a small model of a wind-energy machine.

There is, it is true, a widespread public impression that solar energy is still too expensive for general use, with much of the public failing to distinguish between an expensive, commercially bought and installed active solar hot-water or space-heating system and the, at least now, more economical solar applications such as solar greenhouses, do-it-yourself hot water systems, and passive applications, particularly in new houses. But on the plus side, at least for this region, you hear much less frequently now than, say, a year ago, the question whether solar is not more suited to more sunny areas of the country than New England.

The message that New England is eminently suited now for passive solar systems in new homes and certainly suited for energy-efficient shelter seems to be sinking in, nevertheless; it was surprising to encounter so many people either building new homes using passive solar systems or planning to build them. At the same time the trend to more energy-efficient shelter seems to be gaining momentum, with underground houses holding some interest but even more intense interest in the long-time energy-efficient post-and-beam houses, with builders of these houses backlogged in some cases to over a year's work. Many builders, it is clear, have not understood the energy message and may in this respect be behind the public; one individual told me in almost shocked tones about expensive new homes being built in the woods nearby with no access to the sun.

The current vitality of the alternative energy movement, even with the general disbelief in the "energy crisis," seems due in large measure to three things. One is the success of the national "Sun Day" celebration, which appears to have stimulated an enormous and still continuing interest in solar energy. It seems difficult to overestimate the impact this educational campaign seems to have had upon the public consciousness; schools and libraries responded to the announcement of this event with special energy exhibits, and school children in particular dug into assigned solar projects tied in with "Sun Day" with considerable gusto.

What is also maintaining this vitality is the day-in and day-out activities and educational efforts of conservation and alternative-lifestyle groups and organizations. Every area in New England seems to have one, and although such organizations as Total Environmental Action, Inc. in Harrisville, New Hampshire, and the New Alchemy Institute in Hatchville on Cape Cod are among the better-known groups, there are many more smaller groups, such as Habitat in Belmont, Massachusetts and the New Research Group in Barnstable on Cape Cod, which hold workshops and seminars and offer courses in building solar greenhouses and using wood heat. There is also, of course, the continuing work of the Audubon Society and other environmental groups.

Perhaps most important in getting the solar message across to the general public, many of whom may never have heard of the New Alchemy Institute, has been growing visibility of solar-heated houses and buildings (one of the most widely known solar structures in New England, incidentally, appears to be the solar house erected by the New Hampshire Vocational-Technical College just north of Manchester on heavily traveled Route 93). Everyone seems aware of a solar-heated home or solar-heated homes in their own community or has driven by one, and almost everyone, it seems, mentions a friend or relative who has built or is planning to build a solar-heated home. In Scarborough, Maine, in Brewster on Cape Cod, and near Jaffrey, New Hampshire, to name three locations, I was directed to small solar-heated house developments similar to those springing up throughout New England.

This rising visibility appears to be one of solar energy's most powerful promotional tools and probably has made many people who are planning to build a new home stop and think about using solar energy. This visibility is also a tribute to far-sighted architects and builders (there seems to be at least one of each in every town) who are pioneering solar energy as it is to the government's program to finance construction of solar-heated structures. Mention must also be made of the do-it-yourselfers themselves and the popular press, such as *Popular Science* magazine, which in many cases has aided and abetted these do-it-yourselfers in their energy projects.

Despite signs it is growing, the movement to solar energy has been, for its most fervent believers and advocates, disappointingly slow. One bright hope we heard expressed is that those who have put in solar swimming-pool heaters, which now account for over 80 percent of all sales of solar panels and collectors, will see how well solar energy works and go on to install domestic solar water heaters and, later, solar space-heating systems. But for those who feel the energy crisis requires a change in our way of living, the signs seem even more propitious. Cooking on wood-burning stoves, for example, is in the process of becoming almost as popular as heating with a wood stove, and those who are doing it say food tastes much better cooked over wood than over a gas or electric burner (as those who have cooked a hamburger over a wood fire instead of charcoal, with its gaseous fumes, can attest). As a result, many people who may never have heard of an "alternative lifestyle" are finding unforseen benefits in a change in energy lifestyles and, probably unbeknownst to them, are discovering a common ground with the promoters of a change in the way we live and not just in the way we heat our homes. Many solar greenhouse owners, though not yet the mass movement of those who have bought or are buying wood stoves, are also finding that home-grown vegetables taste better than store-bought produce. So in this, as in, for example, the gradual but strong and continuing move of natural foods to the mainstream of American consumers, the alternative-energy movement is becoming for many Americans even at this early date, though in a small way, a movement also to a new and different lifestyle.

59 Ways to Save Energy

By Bruce Anderson

Bruce Anderson said such nice things about me in his solar thesis for MIT that I knew in a flash that he had good judgment, and he's never disappointed me. Now president of Total Environmental Action and executive editor of Solar Age, *he's also the author of the big one in his field—*The Solar Home Book. *One of those deceptively easy-going, soft-spoken guys who make all sorts of fireworks happen, he is doing better things all the time. I can hardly wait to see where he'll be in five years.*

Here I am, lounging comfortably in a nicely overstuffed chair[1], fur-lined shoes[2] discarded, wool-socked feet[3] waving uncertainly in the air, playing with words and ideas and paper[4] and ink.[5] "Fifty-Nine Ways to Save Energy." What can be said that hasn't already been beaten to death? I conclude, "Nothing."

I stare pensively into space. Silent clashes of battle draw my attention to the window: fierce combat as BTUs muster thousands of successful charges and escape through the leak around the cold steel[6] sash. With equal vigor, millions of photons of sunlight bounce pathetically off the dirty[7]

[1] Heavy, overstuffed furniture is usually warmer and cozier than plastic and metal furniture.
[2] Wear insulated or fur-lined shoes in the winter.
[3] Wear thicker socks (or two pairs) in the winter.
[4] Use paper at least twice. In particular, use back sides of paper already used before.
[5] Use non-disposable pens with ink refills. Remember that every item discarded is discarding resources and energy that went into its production.
[6] Avoid metal window sash; it conducts heat (and cold) too well. Use wood instead.
[7] Keep windows clean for maximum solar heat gain.

chicken-wire glass[8] trying to get in.

Drowsy, in discomfort at the gory battle scene, my eyes shift back to the sterile whitewash of light drowning every nook and cranny of the library—enough light to grow a fine crop of wheat, I dare say. Not much rainfall, though. I look up to the sky of this brightly lit field to rows and rows of fluorescent[9] light fixtures. Ah—there's the rub. Let's remove two of the four bulbs[10] and make the whole room much more pleasant—and save energy.

But what? It's already been done! There are only half as many bulbs lighting the room as was called for in the original design and those that remain are still sufficient to imitate a terrific greenhouse.[11]

I glance over at the only other endangered species in the room—will she give me a knowing glance? No, she only gives her tabletop a dull stare—a fluorescent stare at the papers that absorb a mere fraction of the more than 5,000 watts we share together in the room, about 4,980 more than we need. Small 20-watt bulbs shining on our respective sheets of paper would be more adequate.[12]

The warm room air seems to thicken. My wool socks[3] feel soggy—my longjohned[13] pits cry out for Arid—a few other places need Johnson & Johnson—I wipe my brow and look around for the source of heat. I spot the floor register under the window[14] and rush over to close it. Bending over, my left hand supports me against the cold wall—obviously poorly insulated.[15]

A rush of cool air explodes onto my face and a knowing shock penetrates my bones: air conditioning in the middle of

[8]Use two, even three layers of glass, not just one.

[9]Fluorescent light fixtures are more efficient than incandescent bulbs, but natural lighting uses no energy.

[10]Very often, in large buildings, half the fluorescent bulbs can be removed without significantly reducing lighting levels.

[11]Many (most!) commercial and academic environments have far more light than necessary.

[12]Light is most efficiently used when it shines directly on the item needing the light.

[3]Wear thicker socks (or two pairs) in the winter.

[13]Long underwear significantly improves comfort in cooler environments, enabling room temperatures to be kept lower.

[14]Most buildings are so poorly designed that warm air must be introduced at the windows as a means of improving comfort. But warm air against cold windows means greater heat loss.

[15]Insulate walls (and roofs and floors!).

[16]Many buildings waste energy on excessive lighting and then require air conditioning to rid themselves of the resulting heat.

[17]Electricity is usually an extremely wasteful form of energy use. Avoid its use whenever possible. Again, fluorescent bulbs are more efficient than incandescents.

[18]Individual spaces within buildings, even houses, should be zone controlled. The heat or air conditioning for many spaces can then be turned off when not needed.

[19]So many buildings are designed to be machine controlled. People comfort should be of primary concern when designing buildings.

[20]There should be numerous

winter![16] But of course—the excessive heat from the lights can't get out fast enough, even when it's cold outside. I am jolted by the realization that for each unit of electrical energy used by each bulb, three to four units of fossil fuel energy are burned at the power plant, and only a fraction is actually converted into light, the rest being released as heat![17]

I run to the front desk. "I'm sorry, sir, the thermostat is upstairs. It controls the whole building[18]—and it's locked."[19] And the light switches? There's only one switch for the entire room?[20] As far as I know, it's left on all the time, day and night. Seems they use the light to heat the place.[21] In desperation I race to the window. It's locked too![22] Looking carefully, I notice it hasn't been opened in years.

"Thud." My compatriot's head has fallen to the desk—a dazed unconsciousness replaces her dull, hypnotic stare. I grab my shoes and papers, and flee for the door, my coat trailing on the gray carpet behind me. Lunging for the "fresh" air of the softly lit hallway,[23] I glance to the elevator, but head for the stairs instead.[24] Surely the three flights of stairs are more easily negotiable for me than was the transformation of sunlight into fossil fuel millions of years ago to be exploited today in our power plants to produce the electricity to hoist the elevator! Bursting through the fire escape, I find the din of New York City traffic[25] all the more oppressive in the pollution-muted sunlight.[26] Little wonder I had chosen to be inside.[27] A direct relationship: Less traffic, and less heating and cooling and lighting, produces less pollution and more sunshine.

switches strategically located so that individual lights can easily be turned on (and off!) according to need.

[21]A once common myth—leaving lights on saves energy. Rarely true. Turn them off!

[22]Again, too many buildings are hermetically sealed, ostensibly in the name of keeping mechanical control for the sake of saving energy. Very often, if people can open windows, energy can be saved (and people will be more comfortable!).

[23]I should have helped her, and I feel awful now that I didn't. Sorry. I would have turned the lights off as I left, but there was no switch. At least the hall was sensibly lighted. And it looked and felt so much more comfortable.

[24]Avoid elevators; use the stairs and save energy and improve your health.

[25]Too much needless driving! Walk, bus, bike, or train.

[26]Let's eliminate pollution and let the sunshine help keep us warm.

[27]Working and playing outside means less energy use inside for heat, air conditioning, and lighting.

[28]Ride a bus and leave the driving, and the saved energy (and less pollution) to someone else.

[29]The taxi was driving much too fast. Slow down and live—and save energy, too.

[13]Long underwear significantly improves comfort in cooler environments, enabling room temperatures to be kept lower.

[30]Although I strongly advocate sleeping naked, a few nightclothes save energy, or . . .

[31]Sleep with someone—saves energy.

[32]Tired of heavy layers of blankets? Try down quilts. Mmmm, warm.

I head toward the Port Authority—there's bound to be a bus[28] home.

"Hoonnnk." "Sccreeecchh." A blur of yellow descends down upon me[29]—I jump—but too late and am soaked from the curbside puddle.

I wake up in a sweat. I've been undressed down to my long-johns.[13] Next to me, Angie's warm, naked body[30] undulates peacefully.[31] I slip out from under the down quilt[32] and the flannel sheets[33] to the freshness of 66°F[34] pitch-black air and stumble to the window. The Styrofoam insulating shutter wrapped in beige terrycloth pulls easily away from the window,[35] and I set it against the warm wall[36] filled with the equivalent of eight inches of fiberglass insulation.[37] Starlight streams through the Thermopane[38] wood-frame[6] windows, protected on the outside from the cold by yet another pane of glass, an ordinary storm window.[8] The deciduous trees have lost their leaves for the winter, and through the scraggly hug of the sugar-maple branches the dim horizon forbodes the eternally endless day/night cycle. During the winter we have the benefit of that early-morning heat from the sun, but only finely filtered light manages to squeak through the dense leaves during the summer, keeping us cool.[39]

The stillness[40] of the night is interrupted by Dad's grandfather clock downstairs; it chimes six times.[41] Shall I shower[42] and dress now or wait for Angie? Although our water is heated by a solar collector[43] we just had installed, and although we have changed to water-saving faucets,[44] Angie and I still shower together.[45]

Perhaps a hot cup of tea while I wait.[46]

[33]Flannel sheets (often called "sheet blankets") are the greatest!

[34]Turn your thermostat back as far as possible at night. This is one of the best possible energy-saving efforts you can make! Because this house is so energy efficient, it drops only one or two degrees in temperature at night. Turning thermostats back to 45° or 50°F in ordinary houses is not uncommon and saves considerable energy.

[35]Movable insulation, like Styrofoam covered with cloth, can reduce heat loss through windows to about one-fifth or one-tenth.

[36]A well-insulated wall is warm instead of cold. This saves energy by reducing heat loss, and the warmer wall surfaces permit comfort at lower temperatures, also saving energy.

[37]Many houses today are being built with much thicker walls. Where 3½ inches of fiberglass insulation was once the "standard," today insulation having an insulating value equivalent to 8 inches is often the standard, or should be.

[38]Thermopane windows in combination with storm windows are fast becoming popular. In combination with movable insulating shutters, you can have an extremely energy-efficient window!

[6]Avoid metal window sash; it conducts heat (and cold) too well. Use wood instead.

[8]Use two, even three layers of glass, not just one.

[39]Trees remain the best shading devices known to man. They let the sun in during cold winter months and keep it out during hot summers. The transpiration of their leaves adds to their cooling effect around the house.

[40]An excellent energy-conserving house eliminates or greatly re-

I strike a match and take the refurbished Cape Cod lantern[47] by the handle. As the warm glow and I slip through the den together, Tasha, our fickle seal-point Siamese, snuggles into my overstuffed chair, barely mustering a raised eyebrow. I give a knowing glance to the few drops of wine[48] remaining in the cobalt-glazed mug[49] on the ceramic tiles of the floor[50] next to the neat stack of still-untouched paper.[4] So that's how I got to bed last night!

Without warning, a dull hum of distant motors begins, and suddenly the entire south wall of the den starts to shed its insulating veils, and the early-morning rays of the sun leap across the floor.[50] Small blowers—vacuum-cleaner size—have been told by a simple sensor that it's light outside—time to draw the millions of tiny polystyrene beads from between the two layers of glass spaced four inches apart. During the day, the house is lighted and heated by the sunlight coming through the glass. But at night, the beads fill the space between the two layers and transform the heat-losing window into a well-insulated wall.[51] The thick walls and floors of the house absorb most of the heat;[52] excess heat is circulated by a small fan through the gravel under the floor, preventing the room from overheating, and storing the heat for when the sun doesn't shine for several days.[53] Within a minute, the blowers have finished their chore and the winter-wooded mural covers the entire south wall.

It's been three days since the sun shone. The temperature in the house had reached 73°F, but since then (with 20° weather most of the time outside) it's

duces the need for a large furnace. The sound of the furnace blower is then also eliminated or greatly reduced.

[41]There are many clock designs that require no mechanical energy. Clocks don't use much but it all makes a difference.

[42]On the average, showers use less than half the hot water of a bath. An energy-conserving shower uses even less.

[43]A solar water heater makes good economic sense everywhere, especially if your water is presently being heated by electricity.

[44]Using less hot water saves energy, obviously. Using less cold water also saves energy by reducing the pumping energy required to deliver it to your house. Using less also conserves that very valuable resource, water.

[45]Taking a shower together may not necessarily save energy—although it certainly can. But it sure is fun, and that in itself is enough reason to do it.

[46]Warm liquid always warms the body, if not the soul.

[47]Light a lantern instead of flipping a switch and notice how much more you enjoy light—and save energy!

[48]Wine can help keep you warm too—or make you insensitive to cold. And buy local wines—saves on energy-transportation costs.

[49]A locally made mug saves industrial and transportation energy, and supports the local economy.

[50]Tiled floors absorb and hold heat from the sun.

[4]Use paper at least twice. In particular, use back sides of paper already used before.

[51]Again, movable insulation for the windows. This clever and very effective device is known as Beadwall®, developed by Zomeworks

dropped down to 66°F. Another day without sun and I would have been able to fire up our woodstove.[54] These damned energy-efficient houses, anyway—I hardly ever have the opportunity to experience the penetrating pleasure of wood heat!

As I snuff out the unneeded lantern and reach the kitchen, I smile pleasurably at the ripening tomatoes[55] in the greenhouse[56] just beyond the round oak table. With tomatoes at 85 cents per pound, our greenhouse is doing more than saving energy!

I pour a cup of cold water into a mug,[49] plug in the electric heating coil,[57] and drop it in the water. Outside, the Jacobs[58] is quiet now, but I know with confidence that we have plenty of electricity stored up in our batteries.

The water soon starts bubbling. English Breakfast tea always gets me going in the morning, and I retreat to the den with the hot cup. Tasha's lying in the sun now.[59] I reclaim my chair and settle in again for another try. Let's see now. What can I say about saving energy that hasn't already been said fifty-nine times before?

Corporation in Albuquerque, New Mexico.

[52]All the heat coming through the windows must be stored by heavy material (such as concrete and brick and water) in the house to keep the house from overheating.

[53]Sometimes the heavy materials are not sufficient or are too expensive. An insulated bed of rock or gravel under the floor is ideal for storing the excess heat. Warm room air is blown through the gravel and at night is blown back out to heat the house.

[54]Wood-burning stoves and furnaces are great alternatives to oil, gas, coal, and electricity in many parts of the country. In energy-conserving houses, a good, easy-to-use, efficient wood stove is a pleasure to use.

[55]Grow your own food. It saves growing, harvesting, processing, and transportation energy—and is healthier to eat.

[56]A well-designed greenhouse not only provides solar energy, fresh vegetables, and a pleasant avocation, but also it reduces heat loss by acting as a buffer zone between the house and the cold outdoors.

[49]A locally made mug saves industrial and transportation energy, and support the local economy.

[57]A simple electric coil that dips into the water primarily heats the water instead of the stove burner, the pot, the air, the cup, and the water. If you must use a tea kettle to heat water, heat no more than you intend to use. Use a tea kettle that is as lightweight as possible.

[58]A "Jacobs" is a windmill with a long history (from the early 1930s) of use. Wind power is becoming an increasingly viable means of producing electricity.

[59]Lie in the sun. A great way to stay warm—and healthy!

What a Way to Go!

I wish I knew what the title of this means to you. To me, it means what a wonderful direction in which to be heading, but now that I read those five words I see they can be read in other ways.

"What a way to go!" could also be taken to mean what a long, long way we still have to travel, or even what a great way this is to die. I almost wrote a new title before it dawned on me that every one of those way-to-go meanings applied to this great thing I'm about to write.

If only I could remember what it was. Something to do with the way to go. Ah! Yes: where we're heading; that's it; an architect's view of commerce and industry, past and future, good and bad. In three hundred words or less.

We've all heard about the antiquated building codes that prevent badly needed changes in the construction industry, and we all know that the old-time craftsman has disappeared, and they don't build things like they used to, not to mention those $35-an-hour electricians or plumbers or carpenters (or architects) driving construction costs higher every day. It's a B-P, in other words; a bleak picture, and I haven't even mentioned the energy crisis, which has a construction wallop of its own. Yessir, it looks pretty bad until you think about the way we're heading.

More for less is the idea. The low-energy house will be available soon. So will the 50-mile-a-gallon car, which, as you know, will be very quickly followed by the 100-mile-a-gallon car as the world's remaining gallons continue to dwindle. We're being catapulted up to a level of technology unimagined even a few years ago.

Energy-monitoring computers will soon be available. Wired into your house, they'll squeeze the last ounce of efficiency out of every household appliance. Set the dial to "5 percent less energy this month" and, this month, watch life become 5 percent more interesting. We'll have to reach for every stray sunbeam. Then, maybe once a year, we'll be able to put on all the lights for an hour or so, and invite our friends to share the spectacle.

It doesn't mean hard times. It simply means savoring riches instead of squandering them. It calls for all the ingenuity we stifled when there was energy to burn. A month in a lifeboat is not a highly-recommended experience, but those who've had it have come away dazzled by the stamina of even flabby old bodies like mine, and by the cunning and resourcefulness of the survival mind.

Shortages of various kinds used to be temporary. We suffered through them, or went to war to get—or protect—our rightful share. Now we know better. When shortages are permanent, belt-tightening and wars are useless. We've got us just where we want us, the whole race plunging into an overdue era: sip instead of gulp.

Talk about a business boom! We'll have to replace or improve just about everything in the next 10 or 20 years. Talk about deflation; dollars stretch when you live on less. There'll be inequalities, of course, but with scarcity facing rich and poor alike, every talent in the world will be focused on the problem, with benefits for all.

"Less is more," is what architect Mies Van Der Rohe said about his ultra-simple glass and steel buildings. Now, "less is more" means something else: getting out of the pig-wallow that's made the western world, particularly America, so disgustingly gross; and into the taut, clean pleasure of getting the very most out of everything. What a way to go!

Index